Sea Cucumber Stories

Sea Cucumber Stories explores multispecies entanglements and ocean worldings, from a sea cucumber perspective. Drawing on environmental, multispecies and ocean anthropology, it enlists the sea cucumber to tell stories of how and why the ocean and its creatures matter to life on our blue planet. The chapters draw on multisited, multimodal and multisensory fieldwork in Tanzania, where the sea cucumber is undergoing interesting transformations, from an ocean creature created by God to a commodity in the global seafood market, farmed for export to China. While breaking new ground in multispecies anthropology, this creative book builds on the author's earlier work in digital and literary anthropology. Sea cucumber stories are told from an imaginary as well as a scholarly perspective. The sea cucumber is treated as a fellow creature to draw attention to the wonders of multispecies undersea worlds and the troubles of human exploitation of what is perceived as a marine resource.

Paula Uimonen is Professor of Social Anthropology at Stockholm University, Sweden.

Multispecies Anthropology: New Ethnographies
Series Editors: Rebecca Cassidy and Garry Marvin

Editorial Advisory Board:

Radhika Govindrajan, University of Washington, USA
John Hartigan, University of Texas at Austin, USA
Catherine Hill, Oxford Brookes University, UK
Marianne Lien, University of Oslo, Norway
Piers Locke, University of Canterbury, New Zealand
Laura Ogden, Dartmouth College, USA

Animal Enthusiasms
Life Beyond Cage and Leash in Rural Pakistan
Muhammad A. Kavesh

Human-Horse Relations and the Ethics of Knowing
Rosalie Jones McVey

The Presence of Elephants
Shared lives and landscapes in Assam
Paul G. Keil

Liang Shaoji's Silkworm Art
Untangling Multispecies Craft
Feixuan Xu

The Multispecies Triad of Cattle Ranching
A Horseback Ethnography
Andrea Petitt

Sea Cucumber Stories
Paula Uimonen

https://www.routledge.com/Multispecies-Anthropology/book-series/MANE

Sea Cucumber Stories

Paula Uimonen

First published 2026
by Routledge
4 Park Square, Milton Park, Abingdon, Oxon OX14 4RN

and by Routledge
605 Third Avenue, New York, NY 10158

Routledge is an imprint of the Taylor & Francis Group, an informa business

© 2026 Paula Uimonen

The right of Paula Uimonen to be identified as author of this work has been asserted in accordance with sections 77 and 78 of the Copyright, Designs and Patents Act 1988.

The Open Access version of this book, available at www.taylorfrancis.com, has been made available under a Creative Commons Attribution-Non Commercial-No Derivatives (CC-BY-NC-ND) 4.0 International license.

Any third party material in this book is not included in the OA Creative Commons license, unless indicated otherwise in a credit line to the material. Please direct any permissions enquiries to the original rightsholder.

The open access publication was made possible through the support of a grant from the Swedish Research Council/Development Research. The opinions expressed in this publication are those of the author and do not necessarily reflect the views of the Swedish Research Council.

Trademark notice: Product or corporate names may be trademarks or registered trademarks, and are used only for identification and explanation without intent to infringe.

British Library Cataloguing-in-Publication Data
A catalogue record for this book is available from the British Library

Library of Congress Cataloging-in-Publication Data
Names: Uimonen, Paula author
Title: Sea cucumber stories / Paula Uimonen.
Description: Abingdon, Oxon ; New York, N.Y. : Routledge, 2026. | Series: Multispecies anthropology: new ethnographies | Includes bibliographical references and index. |
Identifiers: LCCN 2025036523 (print) | LCCN 2025036524 (ebook) | ISBN 9781041083634 hardback | ISBN 9781041083658 paperback | ISBN 9781003645054 ebook
Subjects: LCSH: Sea cucumbers--Behavior | Sea cucumbers--Cultures and culture media | Marine animals--Climatic factors | Marine ecology
Classification: LCC QL384.H7 U46 2026 (print) | LCC QL384.H7 (ebook)
LC record available at https://lccn.loc.gov/2025036523
LC ebook record available at https://lccn.loc.gov/2025036524

ISBN: 978-1-041-08363-4 (hbk)
ISBN: 978-1-041-08365-8 (pbk)
ISBN: 978-1-003-64505-4 (ebk)

DOI: 10.4324/9781003645054

Typeset in Times New Roman
by KnowledgeWorks Global Ltd.

Dedication

To the lifegiving ocean and its inspiring creatures.

Contents

List of Figures   *ix*
Acknowledgments   *x*

Introduction: Storytelling with Sea Cucumbers   1
Greetings from the Ocean

Knowing through Imagination and Storytelling with AI 4
Multispecies Storytelling from a Sea Cucumber's Perspective 6
Theorizing with Ocean Creatures across the Indian Ocean 6
Barefoot Methods in Multispecies Ocean Worlds 10
Multimodal and Multisensory Methods in Multisited Fieldwork 16
Narrative Structure and Chapter Outline 18

1 Underwater Creatures   21
  Life on the Seafloor

  Tracing Bêche-de-Mer in Anthropology 24
  Echinoderms and Holothurians in Marine Science 27
  Scientific Gaze and Early History of the Capitalocene 30
  Jongoo Bahari as God's Creation 34
  Species Kin(ds) and Relational Becomings 36
  Stories of Sea Cucumbering 38
    Singing Sjögurka *39*
    Monkeys and Sea Cucumbers in Sama Stories 40
    Creation in Swahili Mythology 40
    Ancient Tales of Sea Cucumbers in Chinese Culture 41
  The Life of a Holothuria Scabra 42

2 Sensory Worldings 44
   We Are Brainless but Not Senseless

   Vernacular Stories of Sea Cucumber Behavior in Coastal Tanzania 46
   Marine Science Descriptions of Holothuria Scabra Behavior 53
   Animal Cognition and Multispecies Correspondence 56
   Sensory Worldings and Sentient Becomings 59
   A Love Story Beneath the Waves 62

3 Suffering Domestication 64
   How Would You Feel About Being Trapped?

   Farming Holothuria Scabra in Ocean Pens 64
   Sea Cucumber and Human Becomings in Artisanal Aquaculture 67
   Multispecies and Multimaterial Ocean Farms 70
   Wild and Farmed Fingerlings for Reproduction 76
   Marine Hatchery for Industrial Production 78
   Risks and Uncertainties in Times of Climate Change 82
   Whispers of the Sandfish 86

4 Blue Expansion 87
   Some Humans Think We Are Marine Gold

   The Tale of the Golden Sea Cucumber 88
   A Deep History of Transoceanic Trade 89
   Postcolonial Exploits of Bêche-de-Mer 92
   Life Stories of Sea Cucumber Fishers and Traders 93
   Neoliberal Expansion and Sea Cucumber Aquaculture 97
   Defiant Resistance to Elite Capture and Dispossession 101
   Survival in the Capitalocene? 105
      Conclusion 106

5 Ocean Creatures Revisited 108
   Voices from the Seafloor

   Sea Cucumber Stories from a Multispecies Perspective 108
   A Message from the Sea Cucumbers: Protect Us, Protect the Ocean 112

   References *113*
   Index *121*

Figures

0.1	Sea cucumber on the seafloor. Photograph by author.	2
0.2	(a–c) Sea cucumbers in a farm. Photographs by author.	3
0.3	Live sea cucumbers in group interview with farmers. Photograph by author.	10
0.4	(a–d) Lobsters and sea cucumbers collected by free-divers. Photographs by author.	12
0.5	(a–e) Walking on the seafloor during fieldwork in multispecies farms in Kaole. Photographs by author.	14
0.6	(a and b) Farmer/divers filming sea cucumbers underwater with GoPro. Photographs by author.	17
1.1	Captain showing us a sea cucumber in the Kaole farm. Photograph by author.	22
1.2	Sea cucumber spitting water when lifted out of the sea. Photograph by author.	23
1.3	(a and b) Drawn illustrations of *Bohadschia* and *Trepang Ananas* (Jaeger, 1833, appendix).	32
2.1	Sea cucumber burrowed into sand. Photograph by author.	46
2.2	(a and b) Abdul showing photographs of raw and processed sea cucumbers on his mobile phone. Photographs by author.	48
3.1	(a–c) Sea cucumber farms of varying sizes with different organizational set up. Photographs by author.	66
3.2	(a–c) Manual maintenance work. Security through CCTV camera or watch tower. Photographs by author.	73
3.3	(a–h) Multispecies sea cucumber farms. Photographs by author.	75
3.4	Entrance to Zanzibar Marine Hatchery. Photograph by author.	78
3.5	Mobile phone showing image of dead sea cucumbers. Photograph by author.	85
4.1	Sea cucumbers being dried for export. Photograph by author.	87
5.1	Sea cucumber in its ocean environment. Photograph by author.	111

Acknowledgments

This book has been inspired by the sea cucumber, a humble ocean creature that lives on the seafloor, taking care of our ocean. Little did I know how significant this creature is for ocean life, so I thank the sea cucumber for the stories it has brought to me. This book has also been inspired by the Indian Ocean in Tanzania. Since my first encounter with this ocean some two decades ago, it has become an important part of my life. It is also the place of my second home, in Bagamoyo. I am ever so grateful to this inspiring coastal environment.

Many humans have also inspired the making of this book. First and foremost, several research assistants in Tanzania, without whose support my fieldwork would not have been possible. Ever since the pre-study, Hussein Masimbi has assisted me, a longtime friend and great research companion. Now a PhD student at University of Dar es Salaam (UDSM), Hussein is putting his ethnographic talent to good use, documenting the cultural heritage of music on the Swahili coast. I have also been assisted by Dr Mary Khatib, a geographer from the State University of Zanzibar (SUZA), and an enthusiastic fieldworker who has coached and facilitated our research, especially on Unguja and Pemba islands. Additionally, Neema Mjengwa, Vitali Maembe and Mussa Sango have helped me out. Importantly, by sharing their stories, the participants in this study have made this book possible. For ethical reasons, they are anonymous in this text, yet their contributions are evident throughout. My gratitude to all of you is profound since there would be no stories to share without you. Asanteni sana.

I have also enjoyed the intellectual support from colleagues at UDSM who joined the research team for *Swahili Ocean Worlds*. Dr Thomas Ndaluka and Dr Rosemarie Mwaipopo from the Department of Sociology and Anthropology have done research on gender-related aspects of peoples' relationships with the ocean. Dr Ronald Ndesanjo, from the Institute of Development Studies (IDS), has studied development challenges related to the Blue Economy. With time, Dr Mary Khatib (SUZA) also carried out her own study for our project, focusing on the inclusion of women in the Blue Economy in Zanzibar. The results of this research are published in a separate publication. The

Acknowledgments   xi

research project *Swahili Ocean Worlds: Fishing Communities and Sea Sustainability in Tanzania* (2022–2024) was funded by the Swedish Research Council/Development Research, grant number 2021-03661. It received ethical clearance from the Swedish Ethical Review Board and research permits from the Tanzania Commission for Science and Technology (COSTECH) and the Zanzibar Research Committee, whose facilitation has been much appreciated.

At Stockholm University, many colleagues have followed and encouraged this research. Rasmus Rodineliussen, fellow ocean anthropologist, has been a constant source of support, along with sensory ethnographer Elin Linder. Both visited my field site in Kaole in conjunction with the Decolonizing Research Methodologies project in November 2023. The BiOrdinary team of Beppe Karlsson, Ivana Macek, Karin Ahlberg, Tomas Cole, Erica von Essen, Gabriel Lennon and Emma Rose Cyr have also been inspirational, not least for paving the way for multispecies ethnography at our department. Literary anthropologist Helena Wulff has always encouraged my creative storytelling. Fellow researchers and teachers, including Eva Maria Hardtmann, Mark Graham, Emy Lindberg, Aina Backman, Emma Dahlin, Renita Thedvall, Molly Sundberg and Raoul Galli contribute to a great working environment, while Anna-Karin Olsson and Peter Skoglund ensure our well-being. Students in social anthropology have followed this research and provided encouraging feedback. So have students in global development, from suggesting an inspiring online music video to baking ginger bread cookies in the shape of sea cucumbers. It is such a privilege to grow our knowledge together, while caring for our planet.

Earlier results have been presented at research seminars and conferences and shared with researchers and networks around the world. At an early stage, during a research seminar at University of Bergen in November 2022, I received valuable feedback from colleagues whose knowledge of the sea cucumber and the ocean in different parts of the world, not least the Pacific, was truly helpful. Special thanks to Geir Henning Presterudstuen, Vigdis Broch-Due, Edvard Hviding and Jon Henrik Ziegler Remme for inspiring insights and inputs, along with all participants at the seminar. I also got valuable feedback from colleagues at Stockholm University at a research seminar in April 2023. Together with Rasmus Rodineliussen, we organized a panel on Caring for Ocean Creatures at the annual conference of the Swedish Anthropological Association (SANT) in Stockholm in April 2023. A follow-up panel at the biannual conference of the European Association of Social Anthropologists (EASA) in June 2024, organized with Edyta Roszko from Chr. Michelsen Institute in Bergen, broadened participation from European ocean anthropologists. The papers of this panel were published in a special issue on Caring for Ocean Creatures in *Anthropology Today* in June 2025, which I co-edited with Rasmus. I have also enjoyed learning with ocean anthropologists in the EASA Anthropology of the Seas network (ANTHSEAS), convened by Marta Gentilucci and Raffaele Maddaluno, and the recently formed Blue Degrowth network, initiated by Borja Nogué-Algueró, Jing Sun, Maria Hadjimichael and Irmak Ertör. It is inspiring to collaborate with this growing number of ocean anthropologists.

At Routledge, Katherine Ong has been a responsive and encouraging editor. I am also grateful to the series editors of *Multispecies Anthropology: New Ethnographies*, Rebecca Cassidy and Garry Marvin, for creating this opportunity to probe human-animal relations.

My Facebook and Instagram friends have shared this journey with sea cucumbers, and some of my best friends have also come to share my fascination with this creature. Especially Susanne Agnäs Benseddik and Susanne Andersson, who also share my love of cats and dogs. Annica Trampus brought me to snorkel in the Adriatic Sea, where I could enjoy an abundance of sea cucumbers, happily living on the seafloor, without human interference. It was a truly inspiring experience of coexistence with our fellow creature.

I thank you all! Any errors or mistakes are my responsibility.

Introduction: Storytelling with Sea Cucumbers

Greetings from the Ocean

Hello there! I'm a sea cucumber, a fascinating marine echinoderm. You might think I'm just a squishy blob on the ocean floor, but I play a vital role in my ecosystem. I help recycle nutrients and keep the seafloor clean by munching on organic debris.

I have a soft, elongated body and can range in color from deep browns to vibrant greens. Some of my relatives can even change color to blend into the environment! I may not be the most glamorous creature, but I have my own unique charm. When threatened, I can eject some of my internal organs to distract predators and make my escape – talk about a dramatic exit! So, next time you think of the ocean, remember that I'm here, quietly contributing to the health of my underwater home.

(https://chatgpt.com/)[1]

Sea cucumber? To most of us, this would be a rather unknown ocean creature. It is not even evident that it is an animal, since cucumber signals a vegetable. Yet this marine organism plays a critical role in the ocean ecosystem. It scavenges on seafloors, feeding on plankton and other miniscule seafood. In so doing, it rejuvenates the ocean. Sometimes it is called an ocean cleaner, to capture its ecological function, while reflecting a growing concern for marine environments and sea sustainability. Among fishing communities on the Swahili coast in Tanzania, the sea cucumber is often referred to as jongoo bahari (sea millipede), although the government has also tried to change its name to tango bahari (sea cucumber), to popularize it as a food product. Here jongoo bahari is known and valued as an export commodity for the Chinese market, speculatively framed as a lucrative product in global trade in bêche-de-mer. Sometimes the sea cucumber is even spoken of as *marine gold*. In Chinese food culture and traditional medicine, the sea cucumber is known as *sea ginseng*, and it has been valued for its medicinal properties for millennia. As for the sea cucumber itself, well it simply lives its life on the seafloor, sensing its way in the undersea world.

Sea Cucumber Stories explores multispecies entanglements and ocean worldings, from a sea cucumber perspective. The book is inspired by the postulation that "it matters what stories tell stories" and that we should start "thinking-with", as in "stories of becoming-with" that include various

DOI: 10.4324/9781003645054-1

This chapter has been made available under a CC-BY-NC-ND 4.0 license.

2 *Sea Cucumber Stories*

Figure 0.1 Sea cucumber on the seafloor. Photograph by author.

"companion species" (Haraway, 2016, pp. 39–40). So, let us make sea cucumber stories matter in storytelling that encompasses living with the ocean. Perhaps we can even think of the sea cucumber as a companion species, a fellow creature, whose relationship with humans can signify how the ocean and its creatures should matter more to all of us.

Like other lifeforms, the sea cucumber is entangled in a web of relations, which can be appreciated as multispecies worldings in a pluriverse. The emphasis on worlding practices of becoming-*with* in multispecies assemblages (Haraway, 2016; Tsing, 2015), brings forth the relational matrix of biosocial becomings (Ingold & Palsson, 2013). For a more holistic planetary perspective, we can appreciate such worldings in the context of an emergent pluriverse, a world of many worlds (Blaser & de la Cadena, 2018; Escobar, 2020; Ingold, 2022). Having explored literary worldmaking by way of water divinity and world literature in terms of a pluriverse of aesthetic worlds (Uimonen, 2020), in this monograph I will explore relational worldmaking with the help of sea cucumbers. More specifically, we will follow sea cucumbers in various entanglements with human history, from early trade to contemporary practices of artisanal aquaculture.

Our sea cucumber stories are anchored on the Swahili coast in Tanzania, but they cross the Indian Ocean all the way to China, thus capturing the spatiotemporal complexity of *Swahili ocean worlds*. Swahili ocean worlds denotes how the ocean features in everyday life on the Swahili coast (Uimonen & Masimbi, 2021). The Swahili coast stretches from Somalia to Mozambique,

Figure 0.2 (a–c) Sea cucumbers in a farm. Photographs by author.

4   *Sea Cucumber Stories*

thus encompassing an area of remarkable cultural diversity, with a deep history of transoceanic exchange. Our recent research project *Swahili Ocean Worlds* (2022–2024) has explored environmental and social sustainability in fishing communities in Tanzania, through a multidisciplinary team of scholars from Sweden and Tanzania. The overall findings of this study of coastal livelihoods and sea sustainability in the context of the blue economy are available in another publication. In this monograph, I will focus on the sea cucumber, which I have followed more closely over the last few years.

Knowing through Imagination and Storytelling with AI

In this storytelling, I will enlist fictitious sea cucumbers, with the help of artificial intelligence (AI). As Haraway pointed out in her discussion of digital world games, stories "belong to storytellers", who share them in "practices of situated worlding" (Haraway, 2016, p. 87). Since this book tells stories of sea cucumbers, I will engage this ocean creature as a fictitious co-writer in the situated worlding of multispecies storytelling. But since the sea cucumber cannot communicate with humans in a way that we understand, I will give it an artificial voice, by way of artificial intelligence. My original idea was to create fictitious sea cucumber characters through creative writing, but I have decided to explore AI instead, because I find it to be an exceptionally creative writing tool. It also enables me to combine my research interests in digital anthropology, literary worldmaking and multispecies ethnography. And this is why, dear reader, I enlist the storytelling powers of AI to give sea cucumbers a voice.

In using a chatbot to tell stories by sea cucumbers, I hope to enhance our imagination. To appraise the value of such creative mediation, let me turn to Ingold, whose reflections on imagination, science and truth take us in interesting directions: "The role of research, then, is to offer an imaginative opening to truth" (Ingold, 2022, p. 334). In terms of situated worlding, we can note his insistence on how knowing "unfolds from the inside of being", so that by "allowing ourselves into their presence", we can reach a point where "the things with which we study begin to tell us how to observe" (Ingold, 2022, p. 335). But how can we know the sea cucumber from the inside? How can the sea cucumber tell us how to observe, let alone understand? Quietly living its life on the seafloor, the sea cucumber embodies an epistemic challenge, which needs a creative solution: "How can ocean inhabitants, ecosystems and dynamics, teach us a lesson in imagination"? (Brugidou & Fabien, 2018, p. 359). Indeed, when expanding their theorizing on the materiality of the ocean from *wet ontology* to *more-than-wet ontology*, scholars have noted the "limit of analytic prose", since understanding the ocean in all its complexity requires "recourse to the imagination" (Peters & Steinberg, 2019, p. 304).

Introduction: Storytelling with Sea Cucumbers 5

By challenging established modes of knowledge production, AI presents new opportunities as well as challenges to anthropology. Having followed Internet development since its early days (Uimonen, 2001), I cannot help but note certain similarities in the polemic hyperbole that now frames AI as a technology that will change everything, for better or for worse. Meanwhile, anthropologists have defined AI as "a collection of interrelated technologies used to solve problems that would otherwise require human cognition" (Bell, 2021, p. 3). I appreciate this emphasis on human cognition, since AI builds upon what is humanly known, while enhancing the repertoire of the knowable, thus offering alternatives and compliments to our knowledge. But I have also become surprised by the creative abilities of AI. I had expected a more mechanical, automated voice, when asking AI to express a sea cucumber's perspective. Instead, I found it to combine facts in a more creative way than I could imagine.

When using AI, I follow academic conventions on referencing, while clarifying how I have used a chatbot. AI-generated sea cucumber stories are clearly marked as indented citations, with a different font and subheadings. In addition to referencing the chatbot web source, in the footnote I explain what prompt I used and when. When experimenting with AI, I have tried out different prompts, to get a response that I find to be factually correct, based on my own knowledge of sea cucumbers. I have also tried out different prompts to get a sufficiently creative story, sometimes adding artistic prods such as fairy tale. I have used the chatbot ChatGPT, developed by OpenAI. This choice was mainly motivated by familiarity with the app from discussions at Stockholm University over the last few years. These deliberations have often been negative, viewing chatbots as a threat to teaching and learning. By contrast, graduate students in digital anthropology have alerted me to the existence of Replika and other AI-generated social phenomena, which has piqued my curiosity. Meanwhile, I have noticed that Swedish research funders allow the use of AI in grant applications, insisting that it must be used in a responsible manner. I have found this academic situation to be rather paradoxical, but instead of remaining puzzled, it has encouraged me to explore AI in a more constructive manner, as a creative writing tool.

By exploring AI for multispecies storytelling, I aim to contribute to creative anthropology. Over the last few years, there has been an upsurge in more creative forms of anthropological writing (Elliott & Culhane, 2017; Murphy & van Roekel, 2024; Rethmann & Wulff, 2023). In my own work, I have explored writing in different genres (Uimonen, 2020). As noted in a recent publication, the aim of creative anthropologies is to defy "imaginaries of academic convention through the cultivation of a *mundus imaginalis* requiring moments of pause, of introspection, and of discomfort" (Murphy & van Roekel, 2024, p. 2). With my background in digital anthropology (Uimonen, 2012; 2016), and literary storytelling (Uimonen, 2020; 2023), I now venture into the terrain of AI to explore a mundus imaginalis of multispecies storytelling.

Multispecies Storytelling from a Sea Cucumber's Perspective

In the quiet depths of the ocean, I, a humble sea cucumber, dwell among the shifting sands and vibrant coral gardens. My body, soft and supple, drapes across the sea floor, blending seamlessly with the environment. Life is a slow dance for me, a gentle ebb and flow guided by the tides.

As I inch along, I can feel the rhythm of the ocean – the whispers of passing fish, the rustle of seaweed, and the occasional thrill of a curious octopus exploring the nooks and crannies of our shared home. Each creature carries a story, and I find solace in their company. [...]

Nearby, a wise old turtle glides by, her shell adorned with barnacles and memories of the ages. She stops to converse, reminding me of the balance we all share in this underwater world. She speaks of the importance of every creature, from the tiniest plankton to the largest whale, and how our fates are intertwined. I feel proud to play my part, recycling nutrients and providing shelter to small critters that call me home.

As night falls, the ocean transforms. Bioluminescent plankton twinkle like stars in the dark, casting a gentle glow. The sounds of the sea shift as nocturnal creatures awaken. I curl up and settle into the sand, content with the stories I carry and the connections I've forged. In this vast ocean, I may be small, but I am a vital thread in the intricate tapestry of life.

And so, I drift into a peaceful slumber, dreaming of currents, companions, and the endless tales that weave us all together in this vibrant marine world.

(https://chatgpt.com/)[2]

Theorizing with Ocean Creatures across the Indian Ocean

In these stories of becoming-with the ocean and ocean creatures, the sea cucumber is the main character, along with humans living in Tanzania on the Swahili coast in the Indian Ocean. While our storytelling gazes underwater, we cannot ignore the interconnectedness of sea and land, especially when it comes to human engagements with sea cucumbers. In brief, this monograph draws on environmental, multispecies and ocean anthropology to explore and explain what the sea cucumber can tell us about worlding on our blue planet. To interrogate human-ocean entanglements, it also uses the lens of political ecology, while trying to think beyond current constraints through biosocial imaginaries and relational interdependences in pluriversal worldings, inspired by African epistemologies. And following the trade in sea cucumbers, the stories also reach across the ocean to China.

As ocean creatures, sea cucumbers foreground *oceanic rather than terrestrial theorizing*. With an anthropological perspective from the ocean that pushes beyond land-centric ecologies (Brugidou & Fabien, 2018; Dua, 2024; Uimonen & Rodineliussen, 2025), this book contributes to the *oceanic turn* in social sciences, or oceanic churn (Helmreich, 2023). Inspired by the notion of using the ocean as a theory machine (Helmreich, 2011), as exemplified by Hofmeyr's work in comparative literature (2010; 2012), the work by Peters and Steinberg on wet ontologies in human geography (Peters & Steinberg, 2019; Steinberg & Peters, 2015), and more recently Jue's work in media studies (2020), this book insists on anchoring its scholarly gaze in the ocean. But unlike previous studies, it is not so much concerned with the ocean as a medium, as an alien space or as a body of water. Rather, anchored on the seafloor, this book adopts a bottom-up oceanic perspective to interrogate what undersea lifeworlds can tell us about oceanic ecosystems, land-sea interdependences and human-ocean interactions.

Focusing on the Indian Ocean, this book aims to contribute to ocean anthropology from an *Africa-centered perspective*. In his recent *Book of Waves*, Helmreich recognizes the need for more southern theory, and he notes "Thinking from the Indian Ocean may re- and de-orient knowledges about how to read oceans and their waves" (2023, p. 30). Although this book is not concerned with waves or wave science, it elaborates on different knowledges about the ocean among coastal peoples in Tanzania, thus re-orienting our scholarly gaze: to *write the world from Africa* (Mbembe & Sarr, 2023). More precisely, this book writes the world from the Indian Ocean in Africa, also known as the Western Indian Ocean. As the literary scholar Hofmeyr has suggested: "the Indian Ocean offers a privileged vantage point from which to track a changing world order" (2010, p. 721). As a method, the Indian Ocean is a complicating sea, not least by bringing forth "transnationalisms within the Global South", in our case between Asia and Africa (Hofmeyr, 2012, p. 589). These transnational interactions can be traced back to early transoceanic trade, a deep history that challenges and dislocates the modernistic understanding of Western colonialism as defining features of planetary worlding. Moreover, paying attention to oceanic relations in Swahili ocean worlds brings forth African epistemologies that can nuance our understanding of human-ocean relations at large (Ikhane & Ukpokolo, 2023). At this particular juncture in world history, thinking with the Indian Ocean may thus add productive inputs to the postulation that "our planet's destiny might be played out in Africa" (Mbembe, 2021, p. 10).

As an ocean creature, the sea cucumber directs attention to *multispecies ethnography*. In their early review, Kirksey & Helmreich defined this emerging field as: "multispecies ethnography centers on how a multitude of organisms' livelihoods shape and are shaped by political, economic, and cultural forces" (2010, p. 545). They attributed a key starting point for the *species turn* in anthropology to Haraway and her seminal *When Species Meet* (2008), emphasizing new relations, or becomings-with, in multispecies contact zones (Kirksey & Helmreich, 2010, p. 546). In her later work,

8   *Sea Cucumber Stories*

Haraway has encouraged scholars to stay with the trouble of living and dying on a damaged earth, focusing on kin making with other critters and other multispecies worlding practices (Haraway, 2016). On a similar note, scholars have elaborated on arts of living on a damaged planet, elaborating on the entangled histories of landscapes and interspecies sociality in the Anthropocene (Tsing et al., 2017, 2024). More specifically, Lien's seminal work on salmon aquaculture is significant for my work on sea cucumber farming, from her appraisal of aquaculture as multispecies assemblages (Lien, 2015; Lien & Law, 2011), to the problem of fluid dispossession in blue frontiers (Lien, 2024). In recent times, multispecies ethnographers have foregrounded more intersectional concerns, focusing on multispecies justice (Chao et al., 2022). While this work has a decolonizing ambition, it proclaims to follow "Western continental philosophy and political theory, while remaining attentive to ways that justice has been twisted by colonialism, capitalism, and racism" (Kirksey & Chao, 2022, p. 5). A more Africa-centered analysis of multispecies worldings can hopefully contribute to the decolonization of multispecies anthropology.

The sea cucumber can also be approached by way of *environmental anthropology*, especially Ingold's seminal work on relational entanglements in the continuous becoming of the world (Ingold, 2000, 2022). Early work on multispecies ethnography noted that "Studies of animals have a long lineage in anthropology" (Kirksey & Helmreich, 2010, p. 549). Ingold traces his own work back to his PhD research among the Sámi and their reindeer in northeastern Finland in the 1970s (Ingold, 2022, pp. 289–292). His phenomenological approach to the study of animals remains quite distinct from the material-semiotic approach that dominates multispecies ethnography, which he has recently critiqued for its appeal to species multiplicity and notions of universal humanity (Ingold, 2022, pp. 305–306). Instead he points to scholars who have elaborated on *hybrid communities*, arguing that all societies are composed of complex entanglements of humans and nonhumans. While recognizing the value of Ingold's critique, I opt for combining rather than confronting the two lines of thought.

In this monograph, I will combine multispecies and environmental anthropology, since my storytelling is inspired by Ingold's elaborations on anthropological studies *with* people and *with* animals (2022, p. 308). In this case, studying *with* coastal people in Tanzania and *with* sea cucumbers. My focus on sea cucumbers is also inspired by the emphasis on arts of noticing and arts of living in multispecies ethnography (Tsing, 2015; Tsing et al., 2017). Not only is this ocean creature under-researched in multispecies anthropology but also since it lives on the seafloor, it remains out of sight for humans. Meanwhile, paying attention to the sea cucumber leads to more careful noticing of its environment, the ocean, not least how seascapes become sites of *extraction* (Kirksey & Chao 2022, p. 3). In this part of the world, we can appreciate how the sea cucumber mediates the art of noticing the "unruly edges" of global capitalism, places where "thinking through precarity" forces us to think beyond Western "dreams of modernization and progress" (Tsing, 2015, p. 20).

Focusing on the sea cucumber, I will combine an *Africa-centered perspective* with a *transoceanic outlook*, for a spatiotemporal appraisal of a deep history of trade and interaction between Tanzania and China. In addition to being an intriguing ocean creature in its own right, the sea cucumber has featured in transoceanic trade for centuries, if not millennia. Valued as bêche-de-mer, the sea cucumber has been collected and processed throughout the Indo-Pacific region for the Asian market, primarily for Chinese customers. Predating the colonial era, this trade offers insights into the deep history of relationships between African and Asian cultures (Anshan, 2010; Kimambo et al., 2017; Qiang, 2010; Shivji, 2006).

As a valued commodity in the global market, sea cucumber trade brings attention to contemporary capitalist configurations, such as blue growth and blue economy, which from a *political ecology* perspective constitute problematic forms of oceanic exploitation. As such, the sea cucumber exemplifies world-ecology in the *Capitalocene* (Moore, 2016a), which in our case offers a more appropriate analytical tool than the Anthropocene (Hornborg, 2020). Earlier work has shown how the rhetoric of triple wins in the blue economy – economic growth, improved livelihoods and environmental protection – conceals exploitative forms of ocean grabbing (Barbesgaard, 2018). Not least by reducing access to and availability of blue commons (Standing, 2023). Indeed, our own research project has shown that the blue economy model tends to magnify existing inequalities (Mwaipopo & Ndaluka, 2023).

When approaching the sea cucumber from a critical lens that reckons with the ecological troubles of global capitalism, it is important to find hope, to *imagine our ocean otherwise*. Recent work on imagination offers theoretical pointers for such efforts, emphasizing the cultural significance of the imagination in the ontogenesis of our world (Ingold, 2022), as well as ontological politics of imagining that another world is possible (Escobar, 2020). As an advocate of pluriversal thinking (Uimonen, 2020), I will use this opportunity to engage nonhuman others to encourage us to imagine life on our blue planet otherwise.

It is worth revisiting Haraway's prompts for storytelling, and her insistence that "we *must* change the story; the story *must* change" (2016, p. 40, emphasis in original). She elaborates on the need to narrate and to think "outside the prick tale of Humans in History", insisting that to "think-with is to stay with the naturalcultural multispecies trouble on earth" (2016, p. 40). Reflecting on Latour's suggestion that we should tell Gaia stories, she proposes geostories as an alternative. Let's submerge such planetary thinking into the sea cucumber's undersea worlds, to help us re-imagine human-ocean entanglements on our blue planet.

Barefoot Methods in Multispecies Ocean Worlds

The sea cucumbers are still alive, lying in the sand. We are sitting on the beach, shaded by some trees. Our group interview on sea cucumber farming attracts several people, men and women of different ages. They are all involved in the farms that we have just visited, within a short walking distance from the beach. One of the women has brought three sea cucumbers, which she has laid on the sand near where we are sitting. I can see some bodily movements, small contractions indicating that they are alive. "This is like a multispecies interview", I think to myself. Suddenly the woman lifts the sea cucumbers, one at a time. She slits their stomachs with a knife and squeezes out their intestines. She buries the innards in the sand and piles the limp bodies on top. After our interview, she takes the gutted sea cucumbers with her.

(Fieldwork in Mtwara, 14 November 2023)

Figure 0.3 Live sea cucumbers in group interview with farmers. Photograph by author.

Introduction: Storytelling with Sea Cucumbers 11

What can fieldwork with sea cucumbers be all about? Anthropologists who practice "multispecies arts of noticing" pay attention to all kinds of nonhuman life, from "charismatic forms of life" to "creatures that are often disregarded" (Kirksey & Chao, 2022, p. 2). When discussing fieldwork methods, they emphasize "slow, bodily, and intimate" forms of "sensing", while using the metaphor of *rubber boots* to capture field-based studies of landscapes and seascapes (Bubandt et al., 2023, p. 8). While I appreciate and agree with this sensory attention to detail, I would also like to bring attention to some blind spots that may affect the art of noticing, while elaborating on *barefoot* methods in coastal environments. Walking on the seafloor to observe sea cucumbers has been a creative way of sensing their multispecies undersea worlds. But these encounters were not obvious from the outset.

It is only when I look back at the photos from our pre-study in Kaole that I realize we had actually seen sea cucumbers already back then, which I had largely ignored. This was on 30 November 2020, when Hussein Masimbi and I had spent some time on the beach, to talk with fishers. My attention was drawn to divers and the lobsters they had caught. As a former scuba diver, I was impressed by their diving gear and practices. They were free-diving, with just a mask, snorkel and fins; no air tanks, no wetsuits. They carried a netted bag and a sharp stick, to catch and collect lobsters from the seafloor. To mark their location in the sea, they used buoys or empty plastic water bottles that could float on the surface. As the divers returned from sea, I observed them lining up their catch of lobsters in the sand, and hanging their netted bags with gear in a make shift shed on the seashore. The lobsters were still alive when a trader weighed them on a digital scale. He paid the divers in cash, according to the size, before removing the lobsters for storage elsewhere, while locating buyers in Dar es Salaam.

It was after observing the lobsters that I noticed sea cucumbers in a bucket on the beach – at the time I did not know what those weird-looking creatures were, or why they were collected. They were sea cucumbers someone explained to us, and they would be sold to Chinese restaurants. I did not think much of it at the time. They looked like something of little value, a weird ocean creature. And I assumed they were sold to Chinese restaurants in Dar es Salaam. Some weeks later I learned that the largest lobsters were exported live to China. I marveled at lobsters being transported all the way to China, from this small community in Kaole. I still did not realize that the trade in sea cucumbers for export to China was even more common.

A year and a half later, sea cucumbers cropped up in angry discussions on our very first day of fieldwork in Kaole. This was on 22 June 2022 and Hussein and I were just starting out to do fieldwork for the *Swahili Ocean Worlds* project. During the pre-study, we had focused on spirituality in relation to fishing and the ocean (Uimonen & Masimbi, 2021). Now we would explore fishing practices more generally, so hanging out on the beach was

12 *Sea Cucumber Stories*

Figure 0.4 (a–d) Lobsters and sea cucumbers collected by free-divers. Photographs by author.

a good place to start. Shortly after we arrived, we learned of some tensions related to the newly established sea cucumber farm. Some members of the farm association had been accused of stealing sea cucumbers and selling them to local traders. This was not legal and the leaders were quite upset about it. They had taken the matter to the police, but it was eventually settled by the families involved, since social relations carried more weight than the law in this community.

This incident of theft and social tension made us pay more attention to sea cucumber farming, which brought forth ongoing transformations in this fishing community. The government had banned the collection of sea cucumbers in 2006 due to reduced stocks, but was now encouraging farming of sea cucumbers, for export to China. The Kaole sea cucumber farm was the first in mainland Tanzania, run by an association of the local Beach Management Unit (BMU). The government had implemented BMUs in all coastal communities, composed of community members to safeguard the marine environment. In Kaole, the BMU set up the sea cucumber farm in 2019, but since it was not running so well, it was reorganized in 2022, just a few months before we started our fieldwork.

The sea cucumber farm, which was within walking distance from the beach, directed our attention to the ocean, especially its bottom. At low tide, a well-trodden path appeared on the seafloor, alongside the mangroves. We could see people walking on the path, moving to and from boats anchored farther out, or to collect various items from the sea, such as fish bait, seashells or sea snails. Used on a daily basis, it connected land and sea in a very peculiar way.

It was a remarkable feeling to walk on the seafloor, barefoot, to visit the sea cucumber farm. Without shoes I could feel the texture of the seafloor, which varied between muddy and sandy. In the muddy parts I felt unsure of my balance and sometimes my feet sank into the sediment. The temperature of the seawater varied according to season and time of day. Attentive to avoid stepping on anything, my gaze was centered on the seafloor. Walking slowly, I paid attention to various creatures: little crabs that scurried along the bottom, tiny fish caught in water puddles or small jellyfish that glistened in the sun.

The sea cucumber farm itself constituted a fascinating multispecies world. During low tide, the fences bordering the farm were visible even from a distance, green plastic mesh fastened on wooden or iron poles. Upon closer scrutiny, all kinds of lifeforms were attached to the fences. Barnacles were growing on the poles and algae would cling on the mesh. The sea cucumbers themselves were not all that visible from above the surface, except when the water was clear and still. They would usually just lie still on the bottom, or burrowed into the sand, only partially visible in low mounds.

Figure 0.5 (a–e) Walking on the seafloor during fieldwork in multispecies farms in Kaole. Photographs by author.

As will be discussed throughout this book, the multispecies worlds of sea cucumber farms connect land and sea, while traversing the ocean through transnational trade. Similar to the nudibranch in Bubandt's research in tidal zones of West Papua, the sea cucumber draws attention to "historical-ecological connections between land and sea" (Bubandt, 2023, p. 184). Inspired by Caribbean island philosophy of tidal dialectics, Bubandt refers to his snorkeling method as *tidalectic ethnography*. As much as tides are significant to my inquiry, I refrain from applying the concept tidalectic ethnography. The coastal environments I focus on are primarily located on mainland Tanzania, which reduces the applicability of archipelagic theorizing. Nor is the colonial history that is so prevalent in Caribbean contexts entirely applicable to the Swahili coast. Here we encounter a much deeper pre-colonial history of transoceanic interactions between African, Arab and Asian worlds. In a similar vein, while pre-colonial trade was directed and mediated by monsoon winds, I will not stretch my analysis of monsoons to methodological *stickiness* (Bhat, 2023), but rather just acknowledge that the monsoon has different names and meanings in different places. In Tanzania, the monsoon is known as kusi (Southeast Monsoon) and kas kazi (Northeast monsoon), signifying the direction of monsoon winds. Similar to how monsoon is "much-more-than-a-season" (Bhat, 2023, p. 200), to coastal people in Tanzania, these terms connote much more than wind direction, since monsoons have a strong bearing on weather conditions and the ocean environment at large. Monsoons also affect sea cucumber farms.

As far as footwear and sea cucumbers are concerned, it is worth noting that an earlier study of sea cucumber collection in Indonesia mentions the use of rubber boots. In her detailed description of how women prepare themselves to collect sea cucumbers at low tide, Osseweijer recalls how they "changed into long-sleeved shirts, headgear and rubber boots or canvas shoes" (2005, p. 62). They walked for a couple of hours in the tidal flats near a coral reef, looking for sea cucumbers through the water surface. Although her study was done before multispecies studies, she notes "spotting some other interesting animals – red and black spotted sea stars, rays and crabs". By comparison, in the book on rubber boots methods, Bubandt mentions that during fieldwork, he would snorkel with "Papuan sea cucumber collectors" as well as other groups of snorkelers, and among Papuans who enter into the "epipelagic world" would be men "to spear fish or collect sea cucumber", while women would stand waist deep during low tide to fish with poles (2023, pp. 174–175). Unfortunately, he does not dwell further on sea cucumbers. Even so, these descriptions of sea cucumber collection indicate a variety of practices, as well as gear. As mentioned earlier, I have typically walked barefoot to sea cucumber farms, and so have most local sea cucumber farmers, except when using boats to reach farms farther out, which has been the case in some locations. In some cases, farmers or other actors involved in farming have worn simple rubber shoes, including a plastic sandal popular among Maasai people. But more often than not, it has been a barefoot venture into farms located in the rather shallow intertidal zone.

Multimodal and Multisensory Methods in Multisited Fieldwork

Through multisited fieldwork, I have gained a comparative understanding of sea cucumber farming in different locations, with the help of various assistants. I have revisited Kaole on many occasions, where Hussein Masimbi is a resident and which is near my home in Bagamoyo. We have also visited the neighboring village Mlingotini, where a privately owned farm was started by an investor from Dar es Salaam, later owned and managed by a local group. The farm had received some assistance from Kaole, mostly advice. The Kaole farm, which was the first on the mainland, had also collaborated with new farms established in Mtwara region in the south, which we visited, along with a new farm in a village in the neighboring Lindi region. We also visited a new farm on Mafia Island. Since sea cucumber farming in Tanzania originated in Zanzibar, we have also done fieldwork on the islands of Unguja and Pemba, visiting sea cucumber farms and meeting with relevant actors, with the help of Dr Mary Khatib. We also visited the main hatchery in Zanzibar. In Pemba, we visited farms in almost all parts of the island, along with relevant government offices.

Using multisensory methods, we have visited the farms in the ocean, either by foot or by boat, depending on the distance from land as well as tidal levels. This immersion into ocean environments has been central to our understanding of the practices involved in sea cucumber farming as well as the underwater habitats of farmed sea cucumbers. On land, we have observed sea cucumbers in different stages of processing, stored in saltwater or being dried. Being able to touch, smell and see the sea cucumber has helped us become more familiar with this peculiar, yet harmless ocean creature.

Multisensory and multimodal research methods can enhance multispecies ethnography in interesting ways. Since one of the guiding ideas is to explore "new modes of sensing and seeing" (Bubandt et al., 2023, p. 5), we can also explore new ways of using digital and visual media, such as film. As MacDougall pointed out in his outline of visual anthropology as a "different *kind* of anthropology", since "certain social phenomena" are "extremely difficult to approach in any other way" (2006, p. 268, emphasis in original). The underwater world is a good example, and visual media can make this world more visible to humans above the surface (Rodineliussen, 2024). Ocean creatures like sea cucumber are another good example, as discussed here. Through film, anthropologists can show "a different experiential world" and the "sensations of different surroundings", thus capturing sensory knowledge (MacDougall, 2006, p. 270). The use of multimodal research methods can also transform research dynamics in compelling ways, not least through more sensory, collaborative and decolonizing approaches to anthropological knowledge production (Westmoreland, 2022).

With a GoPro camera, we have been able to explore the sea cucumber in its undersea habitat. From the very first time that we visited the sea cucumber farm in Kaole, we have brought along a GoPro camera, which has enabled us to film and photograph under the water surface. In this sense, the camera has extended our vision underwater, observing and recording life on and around the seafloor, without our full submersion.

Introduction: Storytelling with Sea Cucumbers 17

Figure 0.6 (a and b) Farmer/divers filming sea cucumbers underwater with GoPro. Photographs by author.

In places where the water has been too deep to walk around in, sea cucumber farmers have helped us film with the GoPro, free-diving in and around farm pens. When we first tried this out, in conjunction with the filming of *Jongoo Bahari* (Sea Cucumber) in Kaole, a member of the sea cucumber association who was a diver helped us get underwater footage.[3] In addition to filming sea cucumbers on the seafloor, the diver took some creative initiative, as he took a selfie and tried to track a colorful small fish. When we returned to the seashore, he was very excited, proudly showing people around us the underwater footage. I realized that the film enabled him to show others what the ocean world looked like, an environment he was very familiar with from diving, but one that was not all that well known to non-divers.

The GoPro has enabled more collaborative research methods, working *with* interlocutors. In his discussion of collaborative ethnographic filmmaking, Gruber (2022) emphasizes the importance of sharing the camera, thus passing over control of the filming process to others. In my earlier explorations of walking with video, I learned that when interlocutors manage the camera, they also take charge of the research process (Uimonen, 2012). When using the GoPro camera, study participants have taken charge of what to film underwater. They have also been able to share their embodied experience of being underwater, as the recording has captured their movements and sounds, such as the sound of surfacing for air, while documenting the visuals and sounds of the undersea world and its creatures.

We have also produced documentary films, together with study participants. In addition to the short *Jongoo Bahari* film (2022), we made *Living with the Ocean* (2024) in Kaole. Both were made in close collaboration with members of the sea cucumber association, who also engage in other activities in the ocean. In the case of *Living with the Ocean*, two members worked as co-directors, together with a professional filmmaker from the TaSUBa arts college in neighboring Bagamoyo. As co-directors, they contributed to the scriptwriting, and they also enlisted and organized the participation of their fellows in various scenes. This resulted in a participatory video, the making of which everyone involved enjoyed. We have also made documentary films for the Swahili Ocean Worlds project at large, which are available on a dedicated YouTube channel.[4] More photographs are available on our Instagram account.[5]

Narrative Structure and Chapter Outline

This monograph contains five chapters: this introduction followed by four empirical chapters, and a short conclusion. Each chapter includes fictitious sea cucumber voices, marked by italicized and indented text, with due references to the AI source in footnotes. The sea cucumber's perspective is also highlighted in the subheading for each chapter.

In each chapter, ethnographic material is combined with analytical discussions that draw on different disciplinary groundings, as well as other sources of knowledge, from art to illustrations. This eclectic patchwork is inspired by the idea of knowledge production through *patchy epistemics* and the *piling* of different knowledge systems (Tsing et al., 2024). It is an approach that is premised on certain commitments: to think in/from place, to embrace a variety of descriptive modes and to cultivate connections across difference (Tsing et al., 2024, p. 193). This book is also committed to place, in this case, coastal places in Tanzania, especially the oceanic places of sea cucumber farms. When it comes to diversity in descriptive modes, this book encompasses ethnographic and AI-generated storytelling, alongside a variety of scientific standpoints. The scholarly analysis draws on different scientific disciplines, especially anthropology and marine biology, but also history, philosophy and political science. These scientific standpoints are combined with the perspectives of people I have engaged with during fieldwork in Tanzania. Their names have been anonymized for ethical reasons. As for the cultivation of connections across difference, you will find some interesting similarities in scholarly accounts of sea cucumbers in history. More importantly, the sea cucumbers stories will hopefully connect you to this remarkable ocean creature.

This introductory chapter **Storytelling with Sea Cucumbers** has clarified the aims and scope of this book, together with some *Greetings from the Ocean* by an imaginary sea cucumber, whose voice has been generated with the help of a chatbot. The chapter has situated the book in relevant scholarship, especially environmental, multispecies and ocean anthropology, while guiding the reader through the theoretical framework that informs the analysis. The use of artificial intelligence (AI) for storytelling from a sea cucumber perspective has been explained in relation to imagination and creative anthropology. The exploration of multimodal, multisensory and multispecies methods during fieldwork has been reflected upon, including collaborative filmmaking and the use of a GoPro camera for underwater filming. The has been concluded with an outline of chapters and narrative structure.

1. Underwater Creatures introduces the sea cucumber in its underwater habitat, while our fellow creature tells stories of *Life on the Seafloor*. The theoretical discussion revolves around how the sea cucumber is known to humans, tracing it in anthropology, marine science and Swahili culture. The notion of species is rethought, alongside the dynamics of scientific classification and epistemic uncertainty. Visual inscriptions and the scientific gaze are discussed in the context of early history of capitalism. The chapter ends with creative renderings of sea cucumbers in different parts of the world, including a song, folktales and mythology, concluding with an AI-generated story from a sea cucumber.

The sea cucumber has no brain, but it knows its environment through its senses, or as our fellow critter explains it: *We are Brainless but not Senseless*. *2. Sensory Worldings* explores the behavior of sea cucumbers, drawing on observations and reflections from people who engage with them in coastal Tanzania. These stories are compared with scientific descriptions, for a broader understanding of some behavioral traits. The theoretical deliberation explores cognition and communication in sea cucumbers, highlighting the significance of the senses in multispecies worldings, while comparing notions of animal sentience in different cultural contexts. A fictitious love story between sea cucumbers concludes the chapter, illustrating their remarkable reproduction skills.

In *3. Suffering Domestication*, the focus turns to aquaculture, exploring partial domestication in sea cucumber farming. Or as the sea cucumber puts it: *How Would You Feel About Being Trapped*? Sea cucumber farms are described in ethnographic detail, alongside accounts from sea cucumber farmers in different coastal locations, to explore different becomings in multispecies and multimaterial ocean farms, while drawing on anthropology of aquaculture and domestication. Special attention is given to the challenge of reproduction, including efforts to develop more science-based forms of procreation in a hatchery. The chapter ends with a discussion of some dangers that sea cucumbers face in captivity, in the hands of human care as well as environmental changes beyond human control. A sea cucumber delivers a poem on the impacts of climate change to conclude the chapter.

Some Humans Think We Are Marine Gold is how a sea cucumber explains their commodification in the global seafood market in *4. Blue Expansion*. The chapter discusses the bêche-de-mer trade in relation to the expansion of the global blue economy. For a historical background, it looks back at the longue durée of transoceanic trade between African, Arab and Asian worlds, which was followed by European colonialism. Postcolonial relations between China and Tanzania frame the discussion on bêche-de-mer trade in more recent times, which is also told through life stories of fishers and traders. The rapid increase in sea cucumber exploitation is discussed in relation to global power dynamics in the expansion of neoliberalism. The problematic results of this blue expansion are also discussed in relation to different forms of local resistance. The chapter concludes with a cautionary tale of the challenges that sea cucumbers face in the Capitalocene.

The concluding *5. Ocean Creatures Revisited,* or *Voices from the Seafloor*, reflects on the importance of including ocean creatures in the storytelling of the ocean in our troubled times. It shows how science and commerce continue to set the agenda for human relations with our companion species, privileging the commercial value of these bottom dwellers instead of their critical role in taking care of the ocean ecosystem. The book concludes with an important message from sea cucumbers, pleading for our protection.

Notes

1 ChatGPT prompt: Introduce the sea cucumber in its own voice. 29 October 2024.
2 ChatGPT prompt: Multispecies storytelling from a sea cucumber's perspective. 2 November 2024.
3 https://www.youtube.com/watch?v=7Ss5lPRFHZA
4 https://www.youtube.com/@swahilioceanworlds7658
5 https://www.instagram.com/swahilioceanworlds

1 Underwater Creatures
Life on the Seafloor

Greetings, curious traveler. I am a sea cucumber, one of the many unsung caretakers of the ocean's vast, mysterious floor. Allow me to share what life is like in my quiet, sediment-filled world.

Here, the seafloor stretches endlessly – a blanket of sand, silt, and the remains of life drifting down from above. Some might call it desolate, but to me, it's a realm of abundance. The currents carry nutrients from distant places, depositing them gently at my feet – or rather, my tube feet. Everything I need is right here.

My day? Simple, yet essential. I crawl slowly across this landscape, sweeping the sand with my feeding tentacles. Each grain I consume carries tiny morsels of decayed plants and animals, the leftover gifts of the ocean's ceaseless cycle of life and death. I digest the goodness, then expel clean, filtered sediment behind me. Yes, I suppose you could call me a recycler. Without me and my kind, the seafloor would choke under the weight of its own detritus.

I am not alone. Around me are other sea cucumbers, brittle stars, and burrowing crabs. They move about their business quietly, just as I do. We don't bother one another. Occasionally, a fish swims overhead, casting a fleeting shadow in the dim blue light that barely reaches these depths. The peace here is profound.

Of course, life has its dangers. Predators sometimes come too close, and when they do, I have a trick – a rather dramatic one, if I may say so. I can expel part of my insides as a distraction. It's messy, yes, but I regenerate what I lose. Adaptability is the key to survival here.

My world is not fast, nor flashy. It's steady and patient, like the pulse of the ocean itself. While others chase, hunt or flee, I remain constant, fulfilling my role in the grand design.

So now you know a little about me and my life. Not glamorous, perhaps, but essential. Here, on the ocean floor, I find purpose in every grain of sand. And that, I think, is enough.

(https://chatgpt.com/)[1]

DOI: 10.4324/9781003645054-2
This chapter has been made available under a CC-BY-NC-ND 4.0 license.

Figure 1.1 Captain showing us a sea cucumber in the Kaole farm. Photograph by author.

The first time I touched a sea cucumber was in the Kaole farm. It was my first visit, and I was assisted by Masimbi. This was on 30 June 2022, our second week of fieldwork. We had tried to visit the farm a few days earlier, but the ladies on guard duty did not want us to come close. Now we came with Captain, who was a member of the management team, so we could enter. When I look back at my fieldnotes, I realize that I did not write how it felt to touch the sea cucumber. I merely noted that when we found some on the bottom, we picked them up and threw them back into the pens. And that Captain picked one up for us so that we could film it. At the time, I had not yet developed an affinity for the sea cucumber; it was just a weird creature, lying on the sea floor. And I was not yet into multispecies ethnography; my focus was on how people dealt with the sea cucumber. Even so, I was curious and wanted to learn more. So, touching the sea cucumber made sense. In retrospect, I can elaborate on how it feels to touch a sea cucumber, which I have done on numerous occasions ever since.

Touching a sea cucumber is somewhat surprising: it feels firm, neither slippery nor squashy. Its texture is quite unlike a fish, which can be quite slippery. It is surprising, because when looking at it, it is easy to anticipate

Figure 1.2 Sea cucumber spitting water when lifted out of the sea. Photograph by author.

something different. I had not expected its skin to be so robust, nor that it would have a somewhat rugged texture. It was also surprising how it would remain immobile when held in my hand – no wriggling or any other bodily movement to get away. So docile. Sometimes it would spit water when lifted out of the sea, as if reminding whoever picked it up that it was out of its element. It belongs on the seafloor, submerged in seawater.

"Whom and what do I touch when I touch", Haraway asks (2008, p. 35). She refers to touching her dog, but we can also ponder what touching other creatures may be all about. In my case, what do I touch when I touch a sea cucumber? Touch is not just a sensory sensation; it is also a matter of connection. As Haraway (2008) argues: "touch ramifies and shapes accountability", adding "Accountability, caring for, being affected and entering into responsibility are not ethical abstractions […] Touch, regard, looking back, becoming with – all these make us responsible in unpredictable ways for which worlds take place" (p. 35). Touching is thus integral to multispecies worldings of becoming-with.

It is not only what you touch that matters but also who the toucher is. To me, touching the sea cucumber stemmed from scholarly curiosity, sensing multispecies worlding in my hands. When a colleague of mine came to visit, the sensory ethnographer Elin Linder (2024), she also touched the sea cucumber, curiously sensing its body. She discovered something hard on its belly. At first, we thought it was a small shell. But it would not come off; it was stuck. We realized that it was probably a tiny foot. To Captain, touching the sea cucumber came from a different standpoint, from someone engaged in farming sea cucumbers in enclosed ocean pens. In touching these creatures, the anthropologists and the aquaculturalist become knotted into the multispecies worldings of the sea cucumber, yet through rather different entanglements, and by extension, different senses of becoming with.

This chapter acquaints you with the sea cucumber from different perspectives, telling stories of multispecies worldings otherwise. It starts with how sea cucumbers have been known in anthropology, especially in the past, focusing on trade in bêche-de-mer. This trade has been discussed in historical analyses of the world system and Pacific regions, thus connecting the sea cucumber with some of the discipline's classic works and forefathers. In marine biology, the sea cucumber is also known for its commercial value as bêche-de-mer, but above all as a marine species. The biological tracing of sea cucumbers goes back to some of the forefathers in natural science. Broadening science with culture, our focus shifts to the Swahili coast in Tanzania, more specifically Kaole. Here the sea cucumber is known as a creature created by God, in an environment that is both material and spiritual. Islamic teachings on the environment offer some clues into these ontological entanglements, along with some prominent features in African epistemology. The chapter concludes with more artistic renderings of the sea cucumber, from a music video in Sweden to folktales in the Philippines and China. And last but not least, the sea cucumber's own story of becoming.

Tracing Bêche-de-Mer in Anthropology

> Another product in demand in China was the sea cucumber (also called trepang or beche-de-mer), which the Chinese valued both as a food and as an aphrodisiac. This product had long been supplied by Indonesian and Philippine sailors, but European traders began to organize the trade themselves. The collecting and processing of sea cucumbers required a great deal of labor. An average-sized establishment might house some 300 people engaged in cleaning and collecting firewood to dry the sea animals. Native laborers were contracted for, working first under their own chiefs and later under European control.
>
> (Wolf, 1982, pp. 258–259).

It is rather telling that an anthropological classic on world history, *Europe and the People without History,* refers to the sea cucumber trade (Wolf, 1982). In his ambitious effort at *global anthropology*, Wolf (1982) positions himself to "follow an imaginary traveller in the year 1400 and depict the world that he might have seen" (p. 24). In his detailed discussion of connections between peoples and places, he traverses the boundaries between Western and non-Western history, showing how "Everywhere in this world of 1400, populations existed in interconnections", not least through networks of trade, which also reached "from East Africa through the Indian Ocean to the Southeast Asian archipelago" (Wolf, 1982, p. 71). It is in his later discussion of European expansion into the Orient that we come across the sea cucumber. Wolf elaborates how Europeans resorted to opium to enter the reluctant market in China, while searching for other goods of interest to the Chinese. In the Pacific, they found a lucrative trade in sandalwood and sea cucumbers being sold to China. Europeans entered this market and traded sea cucumbers for guns, while selling bêche-de-mer to China. Wolf argued that the sea cucumbers for firearms trade also influenced state formation in the Pacific.

Early sea cucumber trade can also be traced to another influential anthropologist, Sahlins (1993), whose perspective differed somewhat from Wolf's. Sahlins argued that "modern academic hawkers of the World System have given too much credit to the trade in European muskets for local sandalwood and bêche-de-mer (sea cucumbers) as the reason for the interrelated developments in warfare, cannibalism, and state formation in nineteenth-century Fiji", and he refers to Wolf's book in his footnote (1993, p. 21). He contended that Fiji had already become a powerful state of its own accord, well before the bêche-de-mer trade of the 1830s and the arrival of muskets. Instead, Sahlins recognized another ocean creature of cultural significance – the whale, more specifically whale teeth.

The focus on sea cucumbers as a commodity in transoceanic trade is telling of how humans have interacted with sea cucumbers for centuries, as storied by anthropologists. It was Reichman's article on sea cucumber trade in Maine (2013) that alerted me to Wolf's and Sahlins' texts, which he mentioned in his overview of earlier studies. While contextualizing his own study in the export of sea cucumbers to Asian markets, Reichman's study focused more on migrant labor than sea cucumbers. Even so, there are interesting historical links between the USA and the Pacific when it comes to sea cucumber trade.

It was not only Europeans who entered the Pacific trade in sea cucumbers, so did Americans. An environmental historian has documented the social biographies of sea cucumbers in the Pacific, tracing connections between Fiji and Nantucket (Massachusetts) in the 1800s (Melillo, 2015). He shows how North American ships hunted whales in the Pacific from the 1790s to mid-1800s. Some of these marine merchants also became heavily involved in the bêche-de-mer trade with China. In addition to sea cucumbers, they traded in dried fish maws and shark fins, which were of no value in Europe or North America but highly valued in China. Melillo recounts how several New England merchants made their fortunes from Fijian sea cucumbers from the late 1820s to the 1840s.

As indicated by the use of the Malay word trepang for bêche-de-mer, the trade in sea cucumbers has been particularly prevalent in the Pacific, as noted in anthropological accounts from this region. Anthropologists have mentioned the collection of sea cucumbers and lobsters in Indonesia (Pauwelussen, 2017), where the practices of sea cucumber collection have been described in great detail (Osseweijer, 2005). Multispecies ethnographer Bubandt (2023, p. 174) has described snorkeling with Papuan sea cucumber collectors, although his focus has been on another ocean creature: the nudibranch/sea slug. These ethnographic descriptions capture a longer history of artisanal sea cucumber collection in the region.

As shown here, for anthropologists, the sea cucumber has primarily been of interest in webs of human relations, rather than as an ocean creature in multispecies ocean worlds. This anthropocentric approach to the sea cucumber is not unlike how most studies of animals have featured in the discipline, from the perspective of human needs and wants. But when human-animal interactions and interdependencies are approached from a more critical ecological perspective, we glean some more disturbing dimensions in human-sea cucumber histories.

Environmental history of bêche-de-mer trade in the Pacific shows its destructive effects. When European and North American merchants got involved in the trade, they transformed local collection into mass production, involving thousands of locals in exploitative labor, and this "relentless sea cucumber harvest severely damaged Fijis' environment" (Melillo, 2015, p. 458). The firewood needed to process sea cucumbers contributed to deforestation. The negative impact on the sea was no less worrisome, as sea cucumber stocks were depleted. Already in 1878, an article in a US magazine warned that "the beche-de-mer has been overfished" (p. 459). But this did not stop European and North American maritime expeditions to continue their greedy harvests of sea cucumbers and other marine goods until the end of the Second World War. Melillo (2015) concludes that the "forceful transformation of locally managed public goods into commodities for private gain—characterize the environmental history of the Pacific World", which he likens to "large-scale economic theft" that offered "minimal returns to Fijians", while destroying their environment (p. 459).

Let us return to Wolf, this time from a multispecies perspective, to recognize "others without history" in the interconnected histories of capitalism (Tsing, 2024, p. 129). Tsing (2024) foregrounds Wolf's analytic use of *articulation* to trace the "history of capitalism as developing through a contingent series of articulations with other modes of production" (p. 130). While Wolf's history writing incorporated non-Western others, multispecies scholars are now incorporating non-human others, to "open the history of capitalism to nonhumans" (Tsing, 2024, p. 131). In our case, we can incorporate sea cucumbers into the history of capitalism.

The articulation of sea cucumber as bêche-de-mer connects Pacific trade with Euro-American capitalist expansion. Building on Wolf, when appraised as "consequential articulations between vernacular local forms, on the one hand, and those of European traders and capitalists", we can appreciate how "Capitalism often changes local ways of being" (Tsing, 2024, p. 134). It should be recalled that the trade in sea cucumbers between Pacific societies like Fiji and China preceded European expansion. In Fiji, the sea cucumber has been known as dri and in Chinese as haishen (sea ginseng) (Melillo, 2015, p. 453). By comparison, the term bêche-de-mer developed much later: dated to 1814, originating from French and Portuguese. The pre-capitalist trade in sea cucumbers differed from the capitalist mode of production in fundamental ways. In addition to harvesting sea cucumbers on a much grander scale, the organization of labor into large camps supervised by foreign managers differed from local ways of doing things, leading to conflictual relations that were made worse by cultural arrogance, since "most visiting mariners were driven by compulsions of commerce and cared little about Fijian traditions and customs" (Melillo, 2015, p. 457). Melillo argues that by paying attention to such *encounters of value*, we can acknowledge "historical acts of cross-cultural violence" that "underwrite the making of markets" in the capitalist world system, along with their "attendant environmental effects" (2015, p. 452).

Returning to Wolf's imaginary traveler in 1400, we can see how the rise of capitalism after 1450 brought new patterns of *environment-making* and *value* systems leading to the Capitalocene (Moore, 2016b, pp. 96–98). Using Moore's (2016b) oceanic metaphors, we can note how the expansion of Euro-American mercantile capitalism into Pacific trade in bêche-de-mer reflected how "Capitalism has always flourished as archipelagos of commodified relations within oceans of uncommodified life-activity" (p. 90). The manual collection and processing of sea cucumbers as objects of trade or supplementary food was replaced with wage labor and mass production, to exploit this odd commodity's value in the Chinese market. The destruction of forests and ocean creatures was of no concern, nor were the lives or lifestyles of supposedly uncivilized others, in this early expansion of racial capitalism.

Echinoderms and Holothurians in Marine Science

The history of sea cucumbers and bêche-de-mer can also be traced in marine science, which recognizes it as a species in marine ecosystems *and* of commercial value. Tracing the becoming of sea cucumbers in marine science takes us to natural history, with its problematic legacy of racist colonialism. While encouraging the decolonization of natural history for the Anthropocene, multispecies anthropologists find themselves "promoting a revitalized natural history, [yet] we do not skirt the fact that the history of natural history remains indelibly implicated in the brutal avarice and racist typologies of colonialism" (Tsing et al., 2024, p. 7). To get past this colonial residue, they identify "ongoing histories of invasion, dispossession, and extraction", while including "a variety of legacies and practices of observation, rather than limiting observation to Western scientists" (Tsing et al., 2024, p. 7). This epistemic diversification can certainly help us revalue the historical trajectories of human engagements with nature around the world. But since natural science continues to enjoy a privileged position in knowledge building, it is worth learning more about the insights and fallacies of this perspective.

The scientific classification of sea cucumbers can be traced to the work of Carl Linneaus and other early European naturalists who accompanied the global exploits of capitalist expansion. In the Western Indian Ocean (WIO), marine life was first investigated by "apostles" of Linneaus in the mid-1700s: "Thanks to the enthusiastic Swedish naturalists of the 18th century, the precedent that exploratory ocean-going vessels should carry a naturalist aboard was established and accepted by the seafaring nations of the time" (Richmond, 2002, p. 43). In the WIO region, these included a Swede named Peter Forskål, who pioneered the collection of marine invertebrates in the Gulf of Aden, English naturalists on Cook's voyage 1768–1771, Charles Darwin on his around-the-world voyage in 1831–1836 and the Challenger Expedition in 1873–1876 (Richmond, 2002, p. 43). Even so, the WIO region remained relatively underinvestigated compared to other parts of the world, well into the 1950s.

European naturalists developed an empiricist natural science whereby Nature was constructed as an abstract. Linneaus' *Systema Naturae* (1735) was the "first step towards a universal natural history", which sought to discover "order *in* nature" (Smethurst, 2012, p. 29). His system of classification marked a "decisive step away from the context of the lived environment towards the abstract space of nature-as-construct" (Smethurst, 2012, p. 30). Plants and specimen were collected around the world and classified into an elaborate taxonomy. Some of these early naturalists traveled along colonial sea voyages; others relied on objects brought back by mariners and other collectors. The notion of organizing nature was part of Europe's civilizing mission, an imperial quest to understand and conquer the world, while "supplanting local with *globalised* local knowledge of nature" (Smethurst, 2012, p. 19, emphasis in original).

A taxonomic introduction to sea cucumbers in East Africa can be gleaned from the aforementioned book, *A Field Guide to the Seashores of Eastern Africa and the Western Indian Ocean Islands* (Richmond, 2002). It is "the first comprehensive field guide to the flora and fauna of the tropical coast of eastern Africa" (Richmond, 2002, foreword). The book explains how in marine science taxonomy, as a marine animal the sea cucumber is categorized in the Echinodermata group, which contains five classes, including sea cucumbers (Holothuroidea), along with featherstars, starfish, brittlestars and sea urchins (Richmond, 2002, p. 300). It underlines that "Echinoderms play an important role in the ecology of shallow water tropical habitats", but they have "little economic value", apart from sea cucumbers, which are processed and traded as bêche-de-mer to Asian markets, throughout the Indo-Pacific region (Richmond, 2002, p. 300). The

Holothuroidea class of sea cucumbers is described as "elongated, soft and often 'leathery' with a ring of tentacles surrounding the mouth. They are commonly seen on tidal flats, in seagrass beds and on coral reefs" (Richmond, 2002, p. 312). The regional name in Kiswahili is listed as "Maji ngoo", which comes across as a somewhat distorted version of jongoo maji, which is how it is known in Tanzania. The book lists three orders of Holothurians, including aspidochirotida, which comprises three families, including the holothuriidae:

> Family **HOLOTHURIIDAE** Small to large, elongated and cylindrical sea cucumbers found from mid-eulittoral to deep waters, often on sand or under boulders. Internally, gonads occur only on left side of dorsal mesentery. Colour patterns often distinctive. Most feed on organic particles mixed in sand which they ingest. At least 30-40 species from five different genera (*Actinopyga, Bohadschia, Pearsonothuria, Labidodemas* and *Holothuria*) have been recorded in the area.
>
> (Richmond, 2002, p. 312, emphasis in original)

I am elaborating on this family since it contains the species *Holothuria scabra*, which is now farmed in Tanzania. It is one of the 34 species listed and described as:

> **H. *(Metriatyla) scabra*** Jaeger Length to 40 cm. Firm, loaf-shaped body. Colour variable, usually uniformly pale cream or grey, often with dark transverse stripes dorsally and white below. Twenty yellowish-brown feeding tentacles. No cuvierian tubules. **Habitat**: in sand, lower eulittoral and deeper. **Distribution**: Indo-Pacific.
>
> (Richmond, 2002, p. 314, emphasis in original)

Let us now dwell on *Holothuria scabra*, which is one of the most researched species of sea cucumber. In a recent publication, *Global Knowledge on the Commercial Sea Cucumber Holothuria Scabra*, a transnational team of marine scientists provide an overview of scientific knowledge (Hamel et al., 2022). Published in the *Advances in Marine Biology* series, the abstract of this 235-page volume underlines that "This review compiles data from over 950 publications" (Hamel et al., 2022, p. 3). It is clearly an impressive publication, with an illuminating rationale:

> The species is important for several reasons: (1) it is widely distributed and historically abundant in several shallow soft-bottom habitats throughout the Indo-Pacific, (2) it has a high commercial value on the Asian markets, where it is mainly sold as a dried product (beche-de-mer) and (3) it is the only tropical holothuroid species that can currently be mass-produced in hatcheries.
>
> (Hamel et al., 2022, p. 3)

30 Sea Cucumber Stories

Similar to how the bêche-de-mer trade was highlighted in the field guide cited above, the introduction to this review of global knowledge even starts with this commercial aspect, through a concise summary of the history and locations of the trade:

> Holothuroids (Echinodermata: Holothuroidea), commonly known as sea cucumbers or holothurians, have been harvested for over 1000 years in the Indo-Pacific regions to supply markets in Asia for seafood and various biological extracts. […] The demand for sea cucumber products has been steadily growing, especially since the re-entry of China into world trade during the 1980s and the explosion of the middle class with its avid consumers.
>
> (Hamel et al., 2022, p. 3)

We can conclude that anthropologists and marine scientists concur in their emphasis on how sea cucumbers matter because of their commercial value, thus foregrounding human relations. These articulations can clearly tell us more about the multispecies history of capitalism, and the role of science in advancing human dominion over nature, which becomes even clearer when appraised in relation to visual representations in science.

Scientific Gaze and Early History of the Capitalocene

When naturalists set out to collect and classify flora and fauna, they used visual methods, not least drawn illustrations. The empirical approach of the Linnean system, which was based on close observation, was part of a *hegemony of vision*, characterized by a "scientific gaze" that was "supposedly disinterested and dis-embodied", claiming "optical truth" (Smethurst, 2012, p. 26). As Latour has argued, in their scientific abstractions, naturalists used words and images to classify nature, what he refers to as *inscription* (1986). An important aspect of this process of inscription was to achieve *optical consistency*, which imbued visual inscriptions with the power of representation (Latour, 1986, p. 13). Building on Latour, in her discussion of skilled vision, Grasseni argues that trained perception and a structured environment have been central in the construction of scientific knowledge (2011, p. 33). She exemplifies this with how naturalists in the 18th century were trained in the skills of drawing before going to the field (Grasseni, 2011, p. 24). But skilled vision is also ideological and embodied, a culturally framed mode of seeing that informs worldviews, learned and enacted in situated practice (Grasseni, 2011, pp. 29–34).

When it comes to drawings of *Holothuria scabra* and related sea cucumbers, they can be traced back to the first naturalist whose name is intertwined with this species: Jaeger. The sea cucumber genus *Holothuria* that this species belongs to has been attributed to none less than Linneaus himself, in the 12th edition of his famous *Systema Naturae* in 1767. Almost seven decades later, the species *Holothuria scabra* was named by Wilhelm Friedrich Jaeger, a German zoologist. His life was shortlived (1812–1834), yet he managed to produce a dissertation in 1833, entitled *De Holothuriis*.[2] In his dissertation, Jaeger mentions that sea cucumbers are found around the world, not least in Asia, where they are also known as *trepang*, prepared by cooking and drying, and sold at a high price, in a chain of commerce between islands (Jaeger, 1833, pp. 28–30). He describes the *Holothuria scabra* in great detail:

Sp[ecies]. 11. Scabra Collect. Schoenl.

Semi pedal or pedal, subcylindrical, with a plain belly, with rounded extremities, sub margined on the side. The back is whitish-grey, but there is a lot of black pigment in the furrows and wrinkles that are most frequent, so that sometimes, especially in contracted individuals, the back appears black. On the belly the skin is thinner, the color gradually decreasing whitish, in some a little reddish. The whole skin is scratched, as if it were covered with sand. Also the feet, which, when found on the back, as well as on the belly, appear rough, hard, and inorganic. Feet 1/2 line long, dark, conical, coming out of black leather perforations. An almost funnel-shaped layer of crenate skin surrounds the circle of tentacles.

(Jaeger, 1833, p. 23, translated from Latin with Google translate, 22 November 2024)

In the appendix of his dissertation, we also find some detailed illustrations (Figure 1.3). These early black and white drawings can be compared with color drawings in the *Field Guide* (Richmond, 2002). Here the *Holothuria scabra* is depicted together with other members of its family. These images are followed by drawings of other species from other families, including the *Thelenota Ananas* (Jaeger), which he had called *Trepang Ananas*. While the names differ somewhat, visually the "Ananas" sea cucumbers are quite similar.

In the latest review of *Holothuria scabra*, we find color photographs as well as drawings and graphic designs (Hamel et al., 2022). The book contains quite a few illustrations, including data graphics, microscopy views and geographic maps.

The development of visual illustrations from drawings to graphic design is instructive of how skilled visions can be appreciated as an *ecology of visual inscriptions* (Grasseni, 2011). Building on Latour's work on the

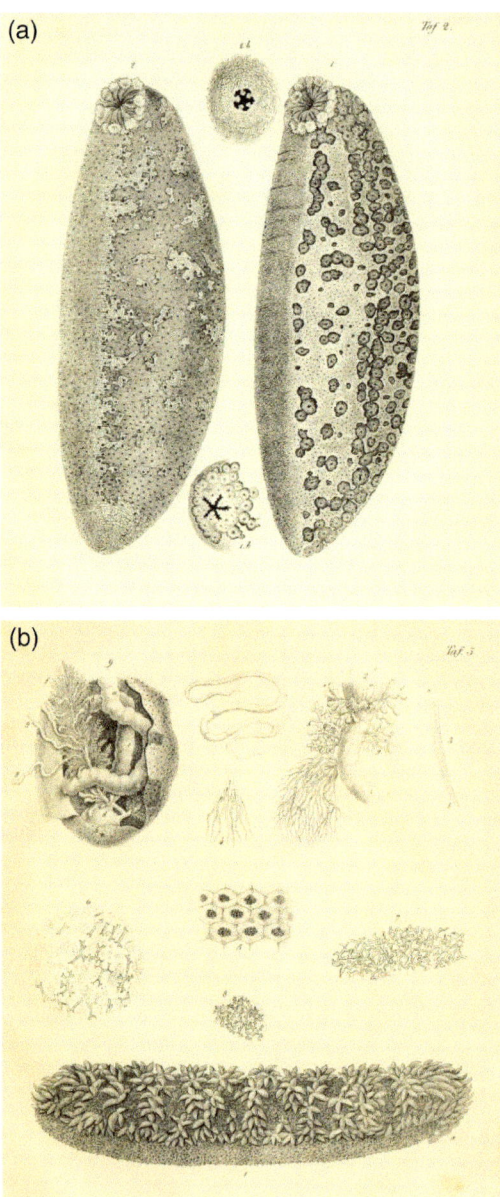

Figure 1.3 (a and b) Drawn illustrations of *Bohadschia* and *Trepang Ananas* (Jaeger, 1833, appendix).

powerful workings of visual inscription in science (Latour, 1986), she suggests that we "consider our visual inscriptions as *artifacts* and that we assess the way in which they contribute to structuring a material, cognitive, and social environment for situated action" (Grasseni, 2011, pp. 42–43). This echoes Latour's focus on visualization and cognition, to further our understanding of how science has enabled the "mobilization and mustering of new resources" through the "manipulation of paper, print, images and so on" (1986, p. 6). After all, powerful economic and political forces have been channeled through the collection and cataloguing of nature: "ideologically charged templates and protocols operated on the texts, images, spectacles and simulacra through which the natural world is represented and re-ordered" (Smethurst, 2012, p. 17).

In the case of marine science, we can note how visual inscriptions remove sea cucumbers from their context, depicting them without their environment. By dislodging sea cucumbers from the sea bottom, they are not only made into scientific objects for human dissection, classification and archiving, but they are also presented as animals without environment, in this case as commercially valuable marine resources.

Through visual inscription, the sea cucumber *Holothuria scabra* becomes objectified, disconnected from its marine ecosystem and re-connected into science as well as expanding maritime trade. Indeed, the whole idea of ordering nature was connected to the capitalist expansion of Euro-American influence. Jaeger was not the only naturalist to recognize the commercial value of sea cucumbers. British naturalist Wallace came across the sea cucumber during his explorations in Singapore, Malaysia and Indonesia, referring to bêche-de-mer as "looking like sausages, which have been rolled in mud and then thrown up the chimney" (Wallace, 1869, p. 435, cited in Melillo, 2015, p. 455). Even though they placed no culinary value on sea cucumbers, European and US merchants were ever so keen to "transform Fiji's natural wealth into commodities suitable for sale to China" (Melillo, 2015, p. 435).

When appreciated as an *articulation* between natural science and colonial conquest in the capitalist world system, we can see how scientific objectification plays into commodification. As Tsing reminds us, "Capitalism as a system depends on stabilizing things as economic resources", which means keeping things under control so that they can be collected and transacted (2024, p. 135). In this case, *H. scabra* is stabilized into a marine resource, which can be collected, processed and sold through maritime trade networks. In this articulation, it is not the culinary or medicinal value of sea ginseng or trepang that is privileged, but rather bêche-de-mer, a thoroughly commodified marine resource that connects Euro-American merchants with markets in China, by way of Pacific intermediaries.

It is important to keep in mind that the scientific gaze did not only classify plants and animals but also classified humans, thus laying the foundation for *racial capitalism*. For instance, the "physical anthropology of Carl Linnaeus (1758)" contained a hierarchical "'classification of seven races, each with associated 'characteristics'" (Farnell, 2011, p. 140). These scientific taxonomies show how the scientific gaze of the time was colonizing, imperialist, and racist (Farnell, 2011; Latour, 1986; Smethurst, 2012). This was by no means limited to natural science, but influenced racist depictions of human races in anthropology as well, including Linnean and Kantian anthropology (Farnell, 2011; Uimonen, 2019).

In terms of capitalism as *world-ecology*, this is also the beginning of *environment-making* in the Capitalocene (Moore, 2016b). As detailed above, the rise of capitalism entailed a new way of organizing nature, but it also brought about new ways of organizing work and life conditions, not least by transforming human activity into labor-power while making humans dependent on cash for their survival (Moore, 2016b, p. 85). This was already evident in how the collection of sea cucumbers in the Pacific was transformed into labor intensive mass production when Euro-American merchants intervened in this transoceanic trade. Recall Melillo's insistence on appreciating bêche-de-mer trade in terms of encounters of value in the making of the world system, the environmental and social consequences of which are still surfacing. But dominant as it may be, this is not the only perspective on sea cucumbers.

Jongoo Bahari as God's Creation

On the Swahili coast in Tanzania, we find a different perspective on the sea cucumber. Here the sea cucumber is entrenched in local understandings of nature that combine Islamic and indigenous worldviews. This ontological hybridity has evolved over centuries of transoceanic exchange and interaction, long before the arrival of capitalist merchants. In more recent times, this hybrid mode of worldmaking has also incorporated capitalist ideals and practices, as will be discussed later on. But for now, let us interrogate the divine powers involved in worldmaking in the material-spiritual coastal environments of Swahili ocean worlds.

In Kaole, the sea cucumber is created by God, like everything else in the world. When we asked people what kind of ocean creature the sea cucumber is, they often referred to God, thus attributing the act of creation to higher powers. As Zuwena, one of the women members of the sea cucumber association told us, after laughing a bit at our question:

> "That creature the way I think of it, first of all it is confusing, because you don't see the eyes, and you see only the mouth and a place for releasing waste. I wonder at the way God has created it, to be like that. It doesn't bite, as if you pick it and it will bite, no. It is just there, you can drag it and it will go, so you can take it anywhere or do anything. I really don't understand how Almighty God created it."

Located on the Swahili coast, Kaole is a predominantly Muslim community and local worldviews are shaped by Islam, as a result of a long history of transoceanic relations. Although most people only complete primary school, they attend madrasa, the Islamic school. For instance, Captain who escorted us to the sea cucumber farm could not study beyond primary school, because his family could not afford it, but he continued with madrasa. According to him, he has learned a lot and can read anything written in the Qur'an. Being Muslim influences how Captain relates to the ocean, not least in his work. As a captain, he is responsible for the boat and its equipment as well as the crew. This work includes a spiritual dimension of prayers and ritual cleansing before fishing, asking for God's protection and blessing, while avoiding to anger the invisible spirits in the ocean. These spirits are also created by God – invisible creatures that exist in the ocean, along with visible creatures, like sea cucumbers (Uimonen & Masimbi, 2021).

To appreciate human-nature relations in this Muslim community, we can explore some foundational principles in Islamic environmental ethics. Scholars refer to the Qur'an and historically formed Muslim practice when discussing nature and ecology in Islam (Ammar, 2019; Saniotis, 2012). Ammar emphasizes *interdependence* and humans as *stewards (Khalifas)*, not owners of the Earth, with humans being part of the ecological community of God's creation, not apart from it (2019, p. 212). Similarly, Saniotis stresses *divine unity in plurality* and *stewardship* as founding ideas of ecological ethics in Islam, with humans as "friends of the earth, not its masters" (2012, pp. 156–157). Both scholars recognize that Muslim practices vary, while alluring to the fragmentation of Islam and the global dominance of development models that have a more utilitarian understanding of nature.

This ontological framing of interdependence differs from the dualist ontology of Euro-American modernity, with its separation of nature/animals and culture/humans. Based on her historical analysis of water beings, Strang has argued that "patriarchal dominion" over nature and non-human living beings can be related to the development of "monotheisms and their beliefs in 'God the Father'" (Strang, 2021, pp. 24–25). While notions of both *dominion* and *stewardship* are found in Judeo-Christian creation narratives, it has been ascertained that the "biblical imperative of 'human dominion' played an important role in the rise of modern science", and the subsequent exploitation of nature (Harrison, 1999, p. 97). This was evident in the work of early naturalists, whose quest for understanding nature was intertwined with ideological claims of human power over nature and European superiority over cultural others, in the expansion of racial capitalism, as discussed above. Meanwhile, although Islam is a monotheistic religion, its emphasis on stewardship is quite different from the Christian emphasis on human dominion, which points to different ontological framings of modernity.

To complicate things further, we also need to appreciate indigenous spirituality in Swahili worldmaking. Scholars of African epistemology underline the spiritual dimension of African ontological reality, a privileging of divine knowledge that tends to be dismissed by Western empiricist science (Chemhuru, 2023). Here we find an "ontological hierarchy of beings", which starts from God, spirits and ancestors, followed by the living, ranked as humans, animals, plants and minerals (Chemhuru, 2023, pp. 94–95). Moreover, each being in this hierarchy has a purpose and influences other beings. Importantly, it is the *spiritual realm* that is the "apex of the African hierarchy of ontology" as well as "the ultimate source of knowledge" (Chemhuru, 2023, p. 95). This emphasis on spirituality is quite different from Judeo-Christian as well as Islamic worldviews, yet they tend to coalesce into various hybrid ontologies in African societies.

This brings us to the relational essence of worldmaking in African cultural contexts, which not only precedes the notion of multispecies worlding but also adds credence to such epistemic advancement in Western science. Scholars of African philosophy have underlined that African epistemology "holds that all things in the universe are unified in a network of relations" and that "humans are involved in a web of relationships with other creatures, including animals, plants, inanimate objects, and even the environment, in this unified, unitary vision of 'what is'" (Ikhane & Ukpolo, 2023, p. 6). This ultimately relational ontology, which underlines the interdependence of all beings, is clearly a very different appraisal of biosocial formation and multispecies worlding than the Cartesian dichotomies underlying the world-ecology and world-economy of capitalist modernity.

Species Kin(ds) and Relational Becomings

I have offered three versions of sea cucumber becomings: as a feature in anthropological studies of the capitalist world system, as a species intricately categorized and studied in marine sciences and as a creature created by God in the Swahili community of Kaole. These are all biosocial becomings, in the sense of "animal becomings in general as the configuration of ensembles of biosocial relations" (Palsson, 2013, pp. 28–29). In becoming sea cucumber, a kind of ocean creature is defined in different ways, depending on which ensemble of biosocial relations it is part of, since becoming is always a matter of becoming *with* (Haraway, 2008). I use the rather fussy "kind of ocean creature" on purpose, because I wish to probe "the concept of species itself" (Ingold, 2022, p. 305). For this, it is worth noting that in the introduction to her seminal *When Species Meet*, Haraway reflects on being schooled as a biologist (she has a PhD in biology), but working as a practitioner of humanities and ethnographic social sciences (2008, p. 13). She notes that "Debates about whether species are earthly organic entities or taxonomic conveniences are coextensive with the discourse we call 'biology'" (2008, p. 17). She adds, "Species is about the dance linking kin and kind", since a biological species is defined according to its ability to interbreed reproductively (2008, p. 17). Let us look more closely at the linking of kins and kinds in marine biology, with its use of kinship terms and other classificatory devices.

As shown in the section above, the classification of sea cucumbers is structured as a hierarchy comprised of class, order, family, genus and species. In the book *Global Knowledge* (Hamel et al., 2022), this categorization differs somewhat from what is specified in the *Field Guide* (Richmond, 2002). In Hamel et.al. (2022, p. 5), the sea cucumber is categorized as follows: class (Holothuroidea), order (Echinodermata), family (Holothuriidae), genus (*Holothuria* Linnaeus, 1767). When it comes to species, it gets more complicated. Firstly, a type species is mentioned, attributed to the scientist who discovered/named it and the year; Type Species: *H. scabra* Jaeger, 1833. Several species are listed under this subgenus, most of them with slight variations on *Holothuria scabra*, such as *Holothuria (Metriatyla) scabra* Jaeger, 1833; *H. scabra* Jaeger, 1833; and *Holothuria (Halodeima) scabra*. This detailed discussion of taxonomy is followed by a table listing common names for *H. scabra* around the world, including: Sandfish (everywhere), Dairo (Fiji), Sand sea cucumber, Hoy sum (Hong Kong), Trepang (Malaysia), Zanga fotsy, Bemavo, Tricot (Madagascar), Jongoo mchanga and Myeupe (Tanzania).

This tedious categorization is actually riddled with taxonomic transformations and uncertainties, which are also known to the scholars themselves. For instance, in the first review of *Holothuria scabra*, the order was listed as "Order ASPIDOCHIROTIDA" (Hamel et al., 2001, p. 133), while in the second review it was listed as "Order Holothuriida (formerly Aspodichirotida)" (Hamel et al. 2022, p. 5). There is no explanation of the name change. Meanwhile, the description remains the same, literally: "This family of about 170 species is typically tropical (Rowe, 1969), although a few species occur in temperate waters. The holothuriids are known as fossils at least as far back as the early Jurassic (Gilliland, 1993)" (Hamel et al., 2001, p. 133; Hamel et al., 2022, p. 5). At the same time, it is also clarified that a possible subspecies, which was included in the first review, turned out to be a different species altogether, as shown by "Allozyme and 16S mtDNA sequence analyses" as well as "a molecular study", so it was omitted from the updated version (Hamel et al. 2022, p. 5).

The making and remaking of species, as well as other taxonomic categories, points to the dynamic fluidity of marine science. To a layman, scientific tests can come across as laborious ways of establishing a new species, the admission of earlier speculative hypothesis having undergone rigorous testing and evaluation to arrive at an evidence-based conclusion. But this is not simply a case of scientists playing the *god trick*, claiming objective, undisputable truths, or making knowledge claims that are "power moves, not moves toward truth" (Haraway, 1998, p. 576). What we can note here is science working towards understanding, not mastery, of marine lifeforms, while recognizing the limitations of what is knowable. It is not unlike the aims of feminist standpoint theorists, as "science becomes the paradigmatic model, not of closure, but of that which is contestable and contested", with the goal of "better accounts of the world, that is, 'science'" (Haraway, 1998, p. 590).

The dynamics of taxonomic categorization are evident in the identification of new species, which has many more dimensions than reproduction alone:

> Determining a new species involves examination of numerous characteristics, including morphological structures, reproduction, developmental biology and evolutionary development (or phylogeny). The examination of the genetic constitution of organisms is becoming a useful tool to this end. Although most of the higher taxonomic groups are unlikely to change, new species are continually being described.
>
> (Richmond, 2002, p. 43).

So, how to rethink species? How to rethink kin and kind? This is quite a challenge for anthropologists, since kinship categories have played a foundational role in our discipline. Remember those kinship diagrams they taught us in undergraduate class? And what to with species now that we are trending multispecies studies?

How about thinking of the sea cucumber as a verb, rather than as a noun? *To be* a sea cucumber is *to* sea cucumber, and sea cucumber*ing* is always a relational act, which happens in the company of other becomings. This is what Ingold suggests, to get past the risk of human centeredness, even in multispecies thinking, emphasising that humans like all animals *occur*, rather than exist, in a ceaseless state of becoming (Ingold, 2022, p. 207). So, humans are always humaning, just like reindeer are always raindeering, to repeat some of his examples. And sea cucumbers are always sea cucumbering? Ingold argues:

> Moreover, the animals, in their animaling, have much to teach us, so long as we allow them into our presence as the beings they are, with their own lives to lead and stories to tell, rather than merely as material-semiotic surrogates for human projects. For at the end of the day, every creature – human or otherwise – answers to the question of what it is by living its own form of life, and in so doing, by contributing to the lives of all the other creatures to which it relates.
>
> (Ingold, 2022, p. 307)

Stories of Sea Cucumbering

Let us now explore creative representations of sea cucumbering though music, folktales and poems, as well as part of God's creation in myths and songs. A song and music video on sea cucumbers by a Swedish artist offers an environmentalist perspective on the sea cucumber. The Sama story on sea cucumbers is a folktale by seafaring people in the Philippines, which is also presented on a website on Sama cultural heritage. In Chinese culture, legends, tales and poems show that sea cucumbers have been appreciated for more than 1,000 years, woven into Chinese art. A folktale is shared here to give an idea of how sea cucumbers are valued in Chinese culture, locally known as sea ginseng (海参;

Haishen). In Swahili culture, the sea cucumber has not yet made an inroad into the arts, but it can be enjoyed as part of God's creation, as described in Swahili mythology. Last but not least, the story that AI tells of life as a *Holothuria scabra* is quite enticing, while corresponding with more scientific accounts.

Singing *Sjögurka*

Enjoy a rap by Kodjo Akolor for Swedish Radio, P3 Gold (2020).[3] The music video is combined with footage from his sing-along performance with a live audience. My translations from Swedish.

"I don't rap about just everything [...] I rap about what is the most meaningful to me. And it may be a bit controversial, but it is *the environment*. And yes, maybe I am a Greta Thunberg with flow, a Greenpeace with rhythm. And I know that Lady Gaga had *A Star is Born*. But, now you will all be introduced to '*A Rapper is Made*'". Kodjo Akolor

Som en sjögurka, rensar havet precis varje dag	Like a sea cucumber, cleans the sea every day
Chillar på botten, det är där jag vill va	Chilling on the bottom, that's where I want to be
Som en sjögurka, som en sjögurka	Like a sea cucumber, like a sea cucumber
Som en sjögurka, rensar havet precis varje dag	Like a sea cucumber, cleans the sea every day
Chillar på botten, det är där jag vill va	Chilling on the bottom, that's where I want to be
Som en sjögurka, som en sjögurka	Like a sea cucumber, like a sea cucumber
Jag vill rensa havets botten	I want to clean the bottom of the sea
För det tycker jag är toppen	Because I think that's great
Näe, sabbar du mitt hav	Look, if you destroy my sea
så kommer jag att blocka dig	then I will block you
Jiddra inte med mitt liv,	Don't mess with my life,
vi kommer återvinna dig	we will recycle you
Ta inte havets liv för givet	Don't take ocean life for granted
Du kommer sabba hela dealen	You will screw up the whole deal
Som en sjögurka rensar havet precis varje dag	Like a sea cucumber, cleans the sea every day
Chillar på botten det är där jag vill va	Chilling on the bottom, that's where I want to be
Som en sjögurka, som en sjögurka	Like a sea cucumber, like a sea cucumber
[x2]	[x2]

Monkeys and Sea Cucumbers in Sama Stories

A folktale on monkeys and sea cucumbers is part of the cultural heritage of the Sama. The folktale has been published as a book that is available through online book stores. The Sama are a seafaring people in the Philippines, and beyond. In their own words:

> *The Bajau, the Badjao, the Samals, and the Sama People*
> The Sama people can be quite hard to classify. Due to the nomadic nature of the Sama they can be found in several countries (especially the Philippines, Malaysia, and Indonesia). In Malaysia they are called Bajau by the Malaysians. In the Philippines, other Filipinos call them Badjaos or Samals, depending on which subgroup of the Sama they belong to.[4]
>
> "**The Monkeys and the Sea Cucumbers**, is a book about conflict. It speaks against bullying and the prevalent mindset that 'might makes right.' This is a generations old fable told by the Sama people of the Sulu archipelago. We are confident in our claims that this is the most fascinating story in the world about sea cucumbers and only a people so closely connected to the sea could come up with such an intriguing folktale".[5]

Creation in Swahili Mythology

Heaven and Earth

> When God's moment had come, He began creating the world of matter. He rolled out the day-sky and the night-sky like an immense tent, or like carpets full of mysterious signs and symbols. In the night-sky He placed the fixed stars like lamps with motionless flames. Others move along the sky, each following a path which only He knows. [...]
>
> He divided the land from the sea, creating the immeasurable ocean on one side, and the high walls of the continents on the other. He heaped up the rocks to be menacing mountains, then told the streams to rush down them in crystal torrents.
>
> He sowed the islands, to be colourful bouquets growing out of the ocean, and a pleasure for the sailing skippers. He commanded quiet pools to reflect the blue skies and the mighty rivers to spread out over the marshes.
>
> He gave a voice to the wind, so that it can whisper as well as roar while it travels over the countries. It pushes the clouds in all directions and it carries the birds on its powerful back. It blows the ships to their destination and it whips up the waves into the frenzy. [...]

Then He told the ocean to be full of fishes of different forms, and so it happened. Only He knows how many there are – and they all have different colours. […]
Only He knows how many species there are. […]
What is there that He has forgotten? Are all these miracles not signs to you of His infinite wisdom, of His immense power?
(Knappert, 1979, pp. 19–22, translated from Swahili mythology)

Ancient Tales of Sea Cucumbers in Chinese Culture

Over a thousand years ago, a couple lived by the sea, and they lived off fishing. The woman gave birth to a boy who the couple called Hai Sheng. The family had to work hard to survive, and lived like this year after year. When Hai Sheng reached 15 years of age, his father developed a strange disease that made him feeble and unable to work. Therefore, the young boy had to take over the fishing. Every day, after fishing, the boy would go to the mountains to gather medicinal herbs in order to treat his father, and would then help his mother with housework. After half a year, the father did not get better. Even worse, the mother developed the same disease. As a result, Hai Sheng had to work day and night.

One day, Hai Sheng wearily rowed his boat out to fish again. After casting his net, he lay in the boat and fell asleep. In his sleep, he saw an old man with a white beard. The old man told Hai Sheng that only a species in the sea could cure his parents. Hai Sheng asked for the name of the creature and what it looked like. The old man smiled and said the species was very ugly, black like a worm, and with four rows of thorns on its body. Finally, the old man told Hai Sheng that the species could be found in Taozi Bay in the Bohai Sea. Before Hai Sheng could say more, he suddenly felt cold, woke up, and realized that he had had a dream.

Just then, the wind started to blow. Hai Sheng drew in the net and got much more fish than usual. After returning home, Hai Sheng told his parents about his dream. After consultation, the family decided to move. They sold everything they had, keeping only the fishing boat and nets, and then they left. Hai Sheng rowed the boat carrying his parents along the shore, asking any passersby if Taozi Bay was along the way. After a year, they finally discovered its location. Once there, Hai Sheng looked around as he walked, and saw a thatched cottage on the beach. As Hai Sheng reached the cottage, an old man emerged from it. Hai Sheng went to the old man and asked him about the creature. The old man replied, "The creature does exist, but nobody here dares to eat it." The next day, Hai Sheng went to sea and captured two specimens. He cooked them for his parents to eat. Every day, Hai Sheng fished more for his parents. After three months, a miracle occurred; his parents got better and could work again. The father asked Hai Sheng how they should call the creature. Hai Sheng replied, "It does not have a name, but I call it seagod (海 神; Haishen) because it lives in the sea, and can relieve people from diseases."

Soon, the emperor heard the story, and dispatched people to bring some seagods back to his palace for him to try. After eating it for a month, the emperor felt very vigorous and said the seagod worked wonderfully, but the name was not proper – "the god" (神; shen) usually refers to an immortal, and the homonym "ginseng" (參; shen) seems more proper. Since then, the seagod (海 神; Haishen) has been called sea ginseng (海參; Haishen).

(*Hai Shen saved his parents* in Yang & Bai, 2015, pp. 2–3)

The Life of a Holothuria Scabra

Ah, the life of a Holothuria scabra. To live as a sandfish sea cucumber is to embrace the slow rhythms of the seabed, an existence of simplicity yet profound importance in the ocean's grand cycle. Let me take you into my world, seen through my tentacle-like feeding structures and felt along my leathery skin.

Morning on the Seabed

As the first rays of sunlight filter through the shallow waters, my day begins, though I lack eyes to see the light. I sense the faint warmth and the gentle currents that swirl over my sandy home. My tube feet – tiny, hydraulic-powered suction cups – press into the soft substrate, propelling me forward in a slow and deliberate march.

Feeding, My Way of Life

The seabed is my buffet. Using my feathery feeding tentacles, I scoop sand and mud into my mouth, savoring the microscopic detritus and organic matter clinging to each grain. It's not glamorous, but it's vital work. As I process this sediment, I digest nutrients that sustain me while leaving behind cleaner, aerated sand. In my slow, patient way, I help keep the seabed healthy, my waste enriching the ecosystem.

Danger Lurks Beneath the Calm

Life isn't without its risks. Crabs and fish eye me as an easy meal, their shadows flickering above. But I have my defenses. My thick, leathery skin is unappetizing, embedded with small, calcified spicules that make me less palatable. If a predator dares to attack, I have a more dramatic strategy: evisceration. Yes, I can expel part of my guts – a sticky, tangled, and sometimes toxic surprise – before slowly regenerating what I've lost. It's a high-stakes gamble, but it works more often than not.

Social Life and Reproduction

Though I often seem solitary, I am not truly alone. I share my home with others of my kind, each of us playing our role in this vibrant underwater community. When the season is right, we gather in larger numbers, synchronized by environmental cues like water temperature and moonlight.

In a grand display, I release my gametes into the water column, mingling them with those of my neighbors. Fertilized eggs drift as plankton, far from this sandy haven. I will never meet my offspring, but I trust they'll settle into the sediment, carrying forward our quiet legacy.

Night on the Seabed

As twilight fades, I continue my unhurried journey across the seabed. Darkness changes little for me; I move with the same steady determination. Occasionally, I pause to burrow into the sand, hiding from nocturnal hunters. The cool grains envelop me, a comforting shield from the unpredictable world above.

My Role in the Great Ocean Cycle

I am more than just a creature of habit; I am an ecosystem engineer. My constant grazing and processing of the seabed help recycle nutrients, feeding the seagrasses and other life forms that depend on them. I play a small but essential role in maintaining the delicate balance of this underwater world.

To live as a sea cucumber, specifically Holothuria scabra, is to embrace simplicity and purpose. My life may seem quiet, unassuming, even insignificant – but every grain of sand I clean, every nutrient I recycle, contributes to the vibrant symphony of the sea.

(https://chatgpt.com/)[6]

Notes

1 ChatGPT prompt: Life on the seafloor as explained by a sea cucumber. 19 November 2024.
2 Available online through Harvard University Library of the Museum of Comparative Zoology https://ia600204.us.archive.org/11/items/deholothuriis00j/deholothuriis00j_bw.pdf
3 https://www.youtube.com/watch?v=LgTq8HJfmJQ accessed on 23 November 2024. Big thanks to the *Fridays for Future* activist and global development student Esmeralda Sjögren for bringing this song to my attention, email exchange on 9 May 2024. And thanks to Kodjo Akolor for permitting me to reproduce the lyrics here (Instagram messages 9-10 May 2025).
4 https://sinama.org/about-sama-people/the-bajau-and-sama-people/
5 https://sinama.org/2021/09/sama-stories-book-2-the-monkeys-and-the-sea-cucumbers/
6 ChatGPT prompt: Tell me about living a sea cucumber life from a *Holothuria scabra* perspective. 23 November 2024.

2 Sensory Worldings

We Are Brainless but Not Senseless

Most of what I know about sea cucumber behavior has been told to me by people who engage with them in coastal communities in Tanzania, as divers, farmers, traders or experts. These accounts offer a good description of how the sea cucumber behaves in its seawater habitat. More specifically, most of these accounts refer to *Holothuria scabra* – the species that is now being farmed. It has also been collected in the wild for many years, alongside other species, so there is quite a lot of accumulated knowledge on its kind.

Vernacular stories are important sources of knowledge on sea cucumbers, their behaviors and habitats, which can be complemented by marine science and other sources. As discussed by Anna Tsing, anthropologists should pay attention to "varied streams of empirical knowledge", including "vernacular natural histories", suggesting that different understandings of landscapes can be appreciated as "different world-making projects" (Tsing in Hastrup et al., 2022, pp. 396–397). To gain a more holistic appraisal of what is going on in such worldmaking projects, she urges anthropologists to "*both* really listen seriously to vernacular understandings of what is going on *and* work with them as part of an intersecting set of histories, including the histories of ecology and natural science" (Tsing in Hastrup et al., 2022, p. 397, emphasis in original). Tsing's holistic approach to different forms of empirical knowledge is also applicable to seascapes and sea creatures.

In this chapter, we will explore the behavior of sea cucumbers through the lens of vernacular knowledge (experiential and expertise), as well as natural science knowledge (marine biology). These different perspectives on sea cucumbers can be appreciated along a spectrum of empirical knowledge, from anecdotal to observational, from hearsay to conviction, from speculative to proven. Sometimes these knowledges intersect and interweave, other times they may be contradictory or counterintuitive. Rather than privileging one form of knowledge over another, I present them as a kaleidoscopic and fragmentary compilation of stories, a plurality of partial perspectives on what the sea cucumber may be all about, as far as some human understandings go.

But first of all, let us probe the sensory worldmaking of the *Holothuria scabra*, aka sandfish, with the help of some AI imagination. One of the

DOI: 10.4324/9781003645054-3

This chapter has been made available under a CC-BY-NC-ND 4.0 license.

peculiar behaviors of this particular sea cucumber is its tendency to bury itself into the sandy bottom. Why is that?

Ah, you wish to understand why I, Holothuria scabra, often bury myself into the sand? Let me share my reasons, from my own humble perspective as a sandfish sea cucumber.

1 Safety First

The ocean is a place of beauty, but also of constant danger. Predators like crabs, fish, and starfish see me as an easy meal. My leathery skin and tough spicules can deter some, but they don't make me invincible. Burrowing into the sand is my way of hiding, a retreat into the safety of the substrate. Here, I am less visible, blending seamlessly into my environment.

2 Rest and Recuperation

Life on the seafloor is demanding. As I tirelessly graze on sediment, my body needs moments of rest. By burying myself, I conserve energy, shield myself from currents, and take refuge from the harsh midday sun filtering through the shallow water. The cool sand envelops me, a natural cocoon.

3 Moisture Retention

Though I live underwater, my leathery body can still lose moisture when exposed to drying currents or fluctuating salinity levels. The sand helps me maintain a stable, hydrated environment around my skin, protecting me from these changes and ensuring I can continue my work.

4 Sediment Stability and Feeding

Sometimes, I burrow not just for protection but for practicality. As I move beneath the sand, I disturb and mix it, bringing buried detritus closer to the surface where I can consume it more efficiently. The act of burrowing helps me access hidden food sources while keeping the seabed healthy for other marine organisms.

5 Avoiding Strong Currents

In shallow waters, strong tidal currents or storm surges can create turbulence. When the water becomes too chaotic, I anchor myself within the substrate. The sand shields me from being swept away or exposed to excessive stress.

6 Biological Rhythms

Some of my kind instinctively bury ourselves at certain times of the day, especially in response to changes in light, temperature, or activity levels of predators. These rhythms are part of my nature, a pattern honed over countless generations to ensure my survival.

Figure 2.1 Sea cucumber burrowed into sand. Photograph by author.

> *To you, burrowing might seem like a retreat, but for me, it's a strategy – a means to survive, thrive, and continue my quiet work of cleaning the seabed. The sand is not just a hiding place; it's my sanctuary, my partner in this slow and steady existence beneath the waves.*
>
> (*https://chatgpt.com/*)[1]

Vernacular Stories of Sea Cucumber Behavior in Coastal Tanzania

"I understand the sea cucumber's behavior", Abdul emphasizes during our interview. We are sitting on the veranda of Hussein's house in Kaole. It is our second day of fieldwork, 23 June 2022, and the Chairperson of the sea cucumber association has advised us to speak with Abdul, to learn more about the sea cucumber farm. Abdul identifies himself as a fisherman. He stopped fishing some ten years earlier, but he has done fishing-related activities all his life, he explains. He has practiced different kinds of fishing, such as diving and net fishing, and he has also worked as an agent, buying and selling fish. His father was also a fisherman. They originated from Mtwara, where they lived near the ocean, and fishing was the main activity. When we interviewed Abdul, he was 43 years old. He was married to one wife, and they had two children. He had primary school education. During our talk, he elaborated on his knowledge of sea cucumber behavior:

"You know you can be fisherman without fishing knowledge. Like myself, I was a fisherman, but in my practice, I like doing research on what I am fishing. That means I know the timing of certain fish. For instance, in which seasons they behave in a certain way, in which season they can be found in a certain location. *I understand the sea cucumber's behavior.* So for those [other fishermen], most of them do not know the procedure for that kind of fishing. I even explained to the Fisheries Officers about the procedure of doing sea cucumber fishing, and they accepted my advice".

During our interview, Abdul shows us photographs of sea cucumbers on his phone, a visual validation of his knowledge claims (Figure 2.2).

Abdul's reflections on what sea cucumbers need and prefer are interwoven in his story of the development of the sea cucumber farm in Kaole. It was initially set up in 2019, with the support of a local businessman, but it was not doing so well. According to Abdul, this was partly due to a lack of knowledge. The location of the pond was not appropriate for the sea cucumbers, so some had escaped.

"They farmed very close to the mangroves, which during the high tide it would look like here [points to the ground]. So, if you keep sea cucumbers like here, where there is no water and they get sunburnt, and later water comes back and goes out again, and comes again in the night, the sea cucumber can leave to a location with water, because they are intelligent to understand where there is deep water. You see, that is what happened".

We asked if the sea cucumber needed to be fed and Abdul explained that cucumbers are intelligent enough to look for their food. They feed on mud and other stuffs in the mud.

"Only mud?", we asked.

"Yes, that's it, nothing else, but it's not just any mud that sea cucumbers eat", Abdul clarified. "That's why I explained, that knowledge, the early pond location it was not right […] sometimes you find the mud they eat are in few places, but where we have relocated nearby that river there are plenty of food they can eat".

He went on to explain that sea cucumbers are very protective; they are smart enough to sense threat, and they can disappear into the mud/sand. By curving his back, he showed with his body language how they would retract when you approached them. They are caught by hand, he explained. At this point, I was impressed by Abdul's knowledge of the sea cucumber, which he

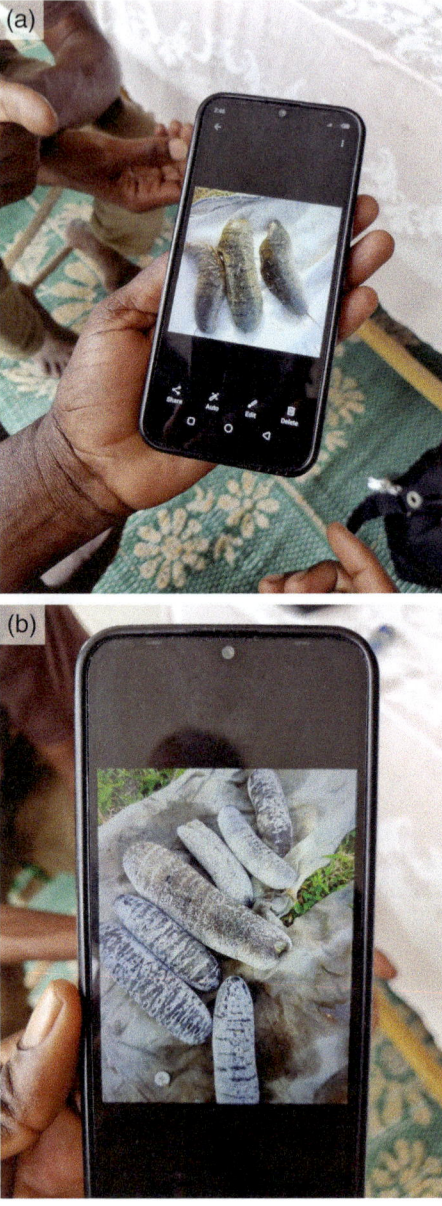

Figure 2.2 (a and b) Abdul showing photographs of raw and processed sea cucumbers on his mobile phone. Photographs by author.

had obviously gained from collecting them as a free-diver. That was years ago, before it was banned, he asserted.

"Do you respect the sea cucumber?" I asked. "Yes, you have to", Abdul retorted.

"Mayai yanatembea!" (the eggs are moving). We are sitting with Neema from the sea cucumber association in Kaole and talking about reproduction. She explains to us:

> When we went to Zanzibar to study [sea cucumber farming], they told us that jongoo do not breed like chicken, like goats or like what. Jongoo breed eggs, that is, they lay eggs. The egg that stays for 3 or 4 days is hatched. When you give birth, you will find a small baby like this [shows the size of a tiny newborn with her fingers]. And the eggs have the ability, that is, when they are laid here, they have the ability to go from here to the mosque [which is located a few hundred meters from her house, where we are sitting]. That is, they have the ability to move a large distance.

When we discuss sea cucumber reproduction, she refers to *mothers* and *children,* thus attributing kinship terms to sea cucumbers and their offspring. She recounts how many thousands of "mothers" they initially planted, and after they had eggs, they had many more. Some of them were also outside the pond, since the eggs had moved. She explained that the eggs could move far: "lina uwezo wa kusafiri likaenda kutotoleka mbali" (it has the ability to travel far away). And this is why you can find sea cucumbers all over the place, also in nearby villages, she reflected: "unakuta bahari yote hii ina jongoo" (You see, this whole ocean has sea cucumbers).

We also talk about sea cucumbers with people in neighboring Mlingotini, a small fishing community far from the main road, with an even slower pace of life than Kaole. Apart from fishing, some villagers have engaged in seaweed farming for several decades, which brings us to the village. To our surprise, we learn that a group has started sea cucumber farming, with some help from the Kaole farm, mostly advise and some fingerlings. But sea cucumbers have been collected in the area for much longer. When visiting the seaweed farm, I observe some men fishing nearby.

> I spotted some men fishing with sticks/rods not far from the [seaweed] farm and we asked what they were fishing for. Samaki [fish], they told us. After a while, one of the men came ashore, and put his net in the sand, since the old man [whom we talked with in the seaweed farm] had shown

interest in buying his catch. We gathered around him, and he proudly displayed what he had caught: squid, crab and jongoo! The creatures were still alive, the squid making a noise, belching in the air when placed in the sand, the jongoo moving slightly in the net. The old man haggled a bit over the price, and bought the 4 squid for only TZS 9,000. It was less than half what they would sell it for in the market. The fisher quickly cleaned them. Pulling out the intestines and the rubbery shield. Some innards were left in a pile, also to be eaten.

I asked about the jongoo. The fisher said the Chinese like them, they can fetch up to 300,000/kilo. His was quite small, he could sell them for 1,000 each. We asked if he eats jongoo and he said yes, it makes you 'joto' [hot]. He was going to bring the *jongoo* home, he explained.

(Fieldnotes from Mlingotini, 6 January 2023).

This fisher was quite unusual in eating jongoo, but similarly to Abdul, he enjoyed the bodily effect of getting hot. This heat is not necessarily related to body temperature, but rather to notions of sexual prowess, since sea cucumbers are known as an aphrodisiac. Another type of seafood that is also known for its aphrodisiacal qualities is a type of sea snail, locally known as *tondo* or *suka*, which women collect in the mangroves.

The sea cucumber is thus incorporated into human sexuality in interesting ways. One of our interlocutors in Zanzibar shared a presentation he had prepared for visitors from Oman, to promote the business of sea cucumber aquaculture. In a slide he had entitled *Scientific Benefits of Sea Cucumber*, he pitched the sea cucumber as *Oceanic Viagra*, with some seemingly copied text explaining how "It has the effect of slowing down the gonad aging, and promoting the erection capacity, so it can consolidate the physical foundation, tonify kidney and improve sperm quality if often have sea cucumbers".

In Pemba, we visited several farms, including one in Muuka-KuuKuu. This was one of the first farms in Pemba, established in 2019. But it has not worked out. All their sea cucumbers had been stolen, before they managed to harvest them. But they were also puzzled by the lack of reproduction. They had waited nine months thinking there would be lots of fingerlings, but there were none. They had even searched in other areas, knowing that the eggs spread with the water. But they didn't find any.

When asked about the ecological function of sea cucumbers they responded: We know there is ecological function because every organism depends on each other, for example where you see the sea cucumber fingerlings also you should see crabs. In those days we saw fingerlings around the mangroves as well as crabs. But nowadays they are rare.

Sensory Worldings 51

In our farms we had fingerlings from different places and some of them were different in shape. I am really curious about these sea cucumbers. When we were farming, I watched the sea cucumbers a lot. I noticed that they stay in groups a lot. I was thinking that maybe they stay like this in groups because they come from different places – each group separates according to where they come from. It also surprised us when it was time to reproduce, they could not reproduce and we kept waiting for them to do so but it was not possible, up until they were stolen.

(Interview in Pemba, 19 July 2023)

It is our first day in Mtwara, having arrived the evening before, after driving for a whole day from Bagamoyo. Hussein has done most of the driving, and I have kept him company, while Mary has rested in the back, having arrived the evening before from Zanzibar. Since time is short, I have decided to fieldwork with both of them to optimize the time available. We are all excited to be in a new location, where none of us have been before. Mtwara is the southernmost region of the mainland Tanzania coast, near Mozambique. After the long drive, we can feel how far away it is. The roads have been good, but this part of the country comes across as sparsely populated, especially when compared to the sprawling city of Dar. It is 13 November 2023, and we spend the next few days visiting various sea cucumber farms in the area, before our journey back, via Lindi and Kilwa Masoko.

Sea cucumbers "live by their senses", he exclaims. They have no brain, nor eyes, they just sense their surroundings, but they do it expertly. They are called ocean cleaners, he tells me, explaining how they clean the water: taking in all kinds of dirt, but what comes out as waste is clean sand.

(Fieldnotes, Mtwara 13 November 2023)

I am walking with Baraka, one of the tutors at the FETA campus in Mtwara. The Fisheries Education and Training Agency (FETA) is a government institute that offers courses on fishing and aquaculture. The Mtwara campus is located on the seaside in Mikindani, a small town a few kilometers from the regional capital of Mtwara. We have just completed a group interview with five tutors. When we are done, they suggest we visit the pens, which we can see through the window, since the tide is low. We can see lots of fences demarcating the pens, the water is very low, in some parts none.

Baraka explains that they bury themselves in the sand, and can survive without water up to 1 hour. They take oxygen from the water. The fences are low. I ask how high the tide gets. He raises an arm above his head – at highest levels I could not even reach the surface, he explains. I ask if sea cucumbers run away, he says no, they stay on the bottom. But I have been told they can bloat and float, I tell him. He says no, they have no respiratory system. Still, I have heard this from several people, so I think he just doesn't know it.

(Fieldnotes, Mtwara 13 November 2023)

Baraka's words stick with me, how the sea cucumbers *"live by their senses"*. It is an apt description of a creature that does not have a brain, yet senses its environment.

These stories exemplify how sea cucumber behavior is known by people involved in fishing or farming them. As you may recall, Abdul mentioned how they shun the sunlight, and to avoid getting burned by the sun, they bury themselves into the sand, or move to deeper waters, when the tide allows. Baraka is convinced that they can survive up to one hour without water, when burrowed in the sand. In Kaole, the pens were moved to deeper water to accommodate the preferences of sea cucumbers; in Mikindani, they are kept close to the FETA office, for ease of human access. Even so, in both places, the fenced-in pens are located in the seawater, so that the sea cucumbers can live on the sea bottom, feeding on whatever is available to them, within the fenced-in area.

The sea cucumber comes across as quite adaptable and resilient, but it is also very sensitive. It certainly knows what kind of environment it prefers, and strives towards it. I am told that as they grow in size, the sea cucumbers seek deeper waters. As mentioned above, they shun the sunlight and can hide themselves in the sand during the day, but at night they come out; they are more visible then, I am told.

Sea cucumbers can swallow a lot of water until they bloat like a balloon, and they can float, moving with the currents. They will do this if they are not happy where they are, for instance, if there is not enough food. I love this story because it shows a surprising mobility in the seemingly sluggish sea cucumber. It also exposes a certain unruliness – if the sea cucumber is not satisfied with its surroundings, it can simply bolt.

The story about floating sea cucumbers shows how varied vernacular knowledge can be. It was first told to me in Kaole, but I heard it repeated in the Zanzibar islands. It was told by people who had seen this action themselves, or knew someone who had seen it. In Mtwara, a FETA tutor denied this ability to float, as noted by Baraka's comment above. It was also denied by the French owner of the lodge where we stayed, who was a scuba diver. He suggested that people had probably just seen sea cucumbers being thrown into the sea and misinterpreted it. It was physically impossible for them to float, he concluded. I was disappointed to hear that, and a bit credulous. How did he know for sure? Could so many people be wrong? When I looked it up online, National Geographic confirmed the floating, as discussed below.

Vernacular understandings of sea cucumbers range from experiential to expert knowledge, acquired in different settings, from embodied observations in the ocean to training in classrooms or workshops. Divers and fishers tend to have limited formal education, typically only primary school education, yet they can be very knowledgeable about sea cucumbers, based on their own experiences of engaging with them in ocean environments. Their skills have been sharpened by a desire and need to know how to best catch sea cucumbers in the wild, when and where to find them and how to approach them. Sea cucumber farmers need to learn how to best take care of sea cucumbers in their ocean pens, so that they may grow. Some people involved in farming have undergone some training, on site in their own locales or in workshops elsewhere. The trainers have acquired their knowledge from different sources and use it as training material, often drawing on global and online sources. The expertise of these trainers varies: some have acquired it through training workshops, while others through formal education in institutes of higher learning. For instance, the FETA tutors in Mtwara have university degrees, often MScs in marine science and/or aquaculture, from universities in Tanzania or elsewhere. But while such expertise tends to enjoy a higher status in local knowledge hierarchies, it does not always concur with experiential knowledge, which in some cases may be more accurate than more academically oriented knowledge. The floating behavior of sea cucumbers is but one example of how academically trained experts may be unaware of what has been empirically observed by people who live with the ocean.

Marine Science Descriptions of Holothuria Scabra Behavior

Let us now explore how sea cucumber behavior is known in marine science. In the review of global knowledge, the behavior of *Holothuria scabra* is described in detail, drawing on extensive studies in different places around the Indo-Pacific region (Hamel et. al., 2022). The descriptions are carefully attributed to different scholars, along with information on where and how the observations were made. Some scholars have observed the sea cucumber in laboratories, others in the field, some in both settings. Their records are carefully summarized and compared, to give a detailed description and analysis of scholarly assessments, with due scholarly rigor.

The *burrowing behavior* is discussed at great length and estimated to be attributable to several factors (Hamel et. al., 2022, pp. 86–92). It is categorized as a daily routine, as described in the chapter *Daily Burrowing Cycle*, which in turn is divided into two parts, *adults* and *juveniles*. The chapter draws on studies conducted in India, Palau, New Caledonia, Torres Strait, Solomon Islands, as well as in laboratory settings. Although methods are not discussed in detail, some interesting observations and experiments emerge in the text. For instance, one scholar used a simple experiment to determine the sensitivity of the anus to light, suggesting that it played a role in the burrowing cycle (Yamanouchi, 1956, in Hamel et.al., 2022, p. 87). He also counted the movements of the anus during various daily activities to observe the sea cucumbers' respiratory rate, which increased when they were burrowed (Yamanouchi, 1956, in Hamel et.al., 2022, p. 89). Observed behavior is seemingly affected by environmental conditions, such as tides, water temperature, water levels, rains, salinity and sunlight. The burrowing behavior is also shown to impact the environment, for instance, the sediment and seagrasses. These behaviors are explained with a degree of *uncertainty*: they are hypothesized, they could be, they are explained as. In general, there is considerable *variation*:

> [F]indings suggest that the circadian behaviours of sea cucumbers can exhibit substantial spatial variation and may be absent or different at certain sites or seasons, and can be mediated by a complexity of factors that vary over short timescales
>
> (Purcell, 2010, cited in Hamel et.al., 2022, p. 90).

Another section focuses on *movement and migration*, with subsections on locomotion and migration patterns, followed by a section on *tagging*. On movement, we learn that sea cucumbers can climb on hard surfaces, such as rocks. And the attentive reader will not be surprised to learn that it can also float:

> *H. scabra* moves chiefly through muscular action, with the help of tube feet densely distributed on the ventral surface of the body wall (see Section 3 for details). It was additionally shown to be able to roll and float using a strategy dubbed *active buoyancy adjustment (ABA)* that helps it travel much faster using currents and changing tides
>
> (Hamel et. al., 2019c, cited in Hamel et al., 2022, p. 43, emphasis added)

This buoyance (ABA) is discussed further, including reasons for it (Hamel et al., 2022, pp. 45–46). It is achieved through a rapid increase in water-to-flesh ratio, which results in bloating and floating, thus detachment from the seafloor. Hamel and colleagues had triggered the action in experimental trials, including the use of decreased salinity and increased water turbidity, and recorded it on video. Additionally, they observed that smaller sea cucumbers could travel quite far:

> Based on video footage from the field, ABA-assisted movements generated speeds of up to several km [kilometers] per day; it was seen to allow small individuals of *H. scabra* to *escape aquaculture pens and drift out at sea*.
>
> (Hamel et al., 2022, pp. 45–46, emphasis added).

These observations of sea cucumber mobility are clearly "challenging the notion of sedentarity in *H. scabra* and other holothuroids" (Hamel et al., 2022, p. 46). Indeed, careful descriptions of how, when and why it moves show that these bottom-feeding ocean creatures get around quite a bit. The section details different speeds and distances that have been measured in adults as well as juveniles, on different surfaces. It is suggested that seagrass beds constitute nursery habitats, with juveniles preferring sand in seagrass before moving on to more open sandy areas or mud flats, while sexually mature specimens tend to migrate to deeper waters, which is also where breeding takes place. Movement also seems to be linked to the availability of food. Again, the analysis is *cautiously worded*: movement patterns appeared to be, they may be linked to, they might spend, it is suggesting that. Even so, graphs are provided to give a more visual and quantifiable presentation of data.

The breeding of sea cucumbers is dealt with at length in a chapter devoted to *sexual reproduction*, with several subsections (Hamel et al., 2022, pp. 52–78). The chapter starts off by asserting that *H. Scabra* has two sexes, although it is impossible to detect on the outside, except during spawning. When it comes to sexual maturity, it is noted that there is some variation on estimates, which is probably a reflection of different methods used. The estimates rely on different criteria: weight, length or age. The overall range for reaching sexual maturity is estimated to be: 18–24 months, or 125–199 mm, or 184–450 grams. Females are expected to have a fecundity of at least one to two million eggs per release. The reproductive cycle varies between studies, although the different stages are generally agreed upon, based on close examination of reproductive organs. As for the life cycle of *H. scabra*, it is not yet well known, but they are estimated to live for up to ten years (Hamel et al., 2022, p. 83).

Reproduction appears to be seasonal, depending on environmental conditions. Environmental factors influence the timing of spawning, especially variations in temperature and salination as well as availability of food. In most regions, there appears to be two peak seasons for spawning, although the actual timing varies from place to place and from year to year. In Tanzania, two peak seasons have been reported, in September as well as between December and January (Kithakeni and Ndaro, 2002, cited in Hamel et al. 2022, p. 69).

In neighboring Kenya, the peaks are reported to occur between October and November, as well as between March and May, during the northeast monsoon season (Muthiga et al., 2013, cited in Hamel et al. 2022, p. 69).

The *moon and tides* also influence spawning, which is a social activity. Before spawning, sea cucumbers tend to aggregate, their social behavior seemingly tied to the lunar cycle, which in turn directs the tide. Interestingly enough, scientists estimate that some form of *communication* takes place, in this case possibly a chemical communication:

> The aggregative behavior of *H. scabra* suggests that chemical communication may be occurring. In the observations from the Solomon Islands, most individuals remained away from each other in the absence of a moon, and started to form pairs, trios and larger groups progressively after the new moon. Aggregation peaked a few days before the full moon when >95% of the individuals participated, and subsequently decreased until the next new moon. Most of the time, pair formations were more common than larger aggregations, with no progressive pattern or correlation to sex.
>
> (Hamel et al. 2022, p. 77)

The spawning ritual is quite intriguing, as male sea cucumbers inspire females to release their eggs by squirting sperm into the ocean. Before spawning, there is some rolling and twisting on the seafloor, with rhythmic contractions of the body. Males then raise their front body part and sway from side to side while releasing their sperm, in a continuous stream. After a short while, females raise their bodies and squirt eggs, in repeated spurts. The spawning can last for a couple of hours, during which males stay erect, while females lie down between their recurrent spurts.

Sexual reproduction is visualized in the *Field Guide* (Richmond, 2002, p. 301). Here the sexual reproduction process is depicted in a collection of drawings showing what holothurians look like inside and out, along with examples of their internal skeleton (ossicles) and their tentacles. The drawing shows how the male ejects sperm and the female eggs, which after fertilization, develop into tiny embryos and later juveniles.

Animal Cognition and Multispecies Correspondence

How do sea cucumbers know and communicate their world? This question points to the epistemic challenge of grasping multispecies worlding with and through companion species, while challenging the hegemony of human exceptionalism. The notion that "conscious thought was the defining characteristic of humans" has dominated much scientific, philosophical and religious thinking, with "cognition (knowing)" usually associated with "consciousness, symbolic reasoning, abstract thought, verbal language" and so on (Mbembe, 2021, p. 86). But all this is now being challenged.

Before exploring cognition in animals, it is worth recognizing that the notion of human exceptionalism is well established in many epistemologies, well beyond Christian/Western worldviews. In his classic Islamic history of the world, *The Muqaddimah,* Ibn Khaldûn argued that "what distinguishes man from the other animals" are "certain qualities peculiar to him", including sciences and craft resulting from his ability to think as well as civilization by dwelling in common in cities and hamlets and co-operation to make a living (Khaldûn, 1967, p. 42). While categorizing humans as animals, they are thus distinguished from other animals, not least through their capacity to think and reflect but also their social skills. Khaldûn was far from alone in this understanding of human exceptionalism. In African ontology, the human ability to both "acquire knowledge" and have "true knowledge" is what makes humans "unique and different from other non-human beings" (Chemhuru, 2023, p. 102). Indeed, it is humans' capacity to comprehend knowledge and truth that is "the principal criterion for differentiating human beings from the lower animals" (Dzobo, 2010, cited in Chemhuru 2023, p. 102).

A more-than-human appreciation of cognition and communication can open our minds to the intricate entanglements of multispecies worldings. When cognition is understood as "engagement of all life-forms with their environment", we can appreciate that cognition is something much broader and that it can extend to plants, micro-organisms and beyond (Mbembe, 2021, p. 86). Mbembe draws on Kathryn Hayles, who discusses cognition in relation to complex technical systems, arguing that "there are nonconscious forms of cognition", which means that cognition is "not limited to humans and life-forms" (Hayles, 2016, in Mbembe, 2021, p. 86). As for communication, although our knowledge of animal communication is still fragmentary, it has been noted that "many animals use an active collective kinesthetic semiotics, as well as chemosensory, visual, and tactile language" (Haraway, 2016, p. 122). Remember what marine biologists suggested about sea cucumbers' chemical communication in relation to their mating rituals under the moon. Communication can clearly take many forms.

When it comes to cognition and communication in sea cucumbers, we are dealing with a living organism. However, one without brain, or a central nervous system. Yet, certainly with cognition. How else could it live in its environment?

Using Ingold's concept of *correspondence,* we can appreciate how sea cucumbers engage with their environment, interacting with others of their own kind as well as in various "cross-species entwinings" (Ingold, 2022, p. 302).

Building on the work of Dominique Lestel, Ingold argues that humans and animals coexist in *hybrid communities*, since all communities of animate beings are hybrid in terms of species composition. He underlines the dynamic essence of these hybrid communities, since the "animal subject is not a bounded entity, set over and against others of its kind, but just one trail of growth and development in a heterogenous field of interests and affects". This captures the correspondence of a world in becoming, in which parts correspond with one another, interweaving in the ongoing life as a whole (Ingold, 2022, p. 349).

In the case of the sea cucumber, it also constitutes a hybrid community of its own, a *symbiotic multispecies community*. Marine biologists have shown that sea cucumbers interact with many symbiotic organisms who can benefit from the association in terms of food or protection (Hamel et al. 2022, p. 122). Sea cucumbers have few predators, although larvae and juveniles risk being eaten by amphipods and copepods, or fish, crabs and sea stars. But adult sea cucumbers can form symbiotic communities with a variety of species. In marine science, these symbiotic relations are categorized according to the impact on the host (the larger organism) by the symbiont (the smaller organism) in terms of mutualism (both benefit), commensalism (no impact on host) and parasitism (negative impact on host) (Hamel et al., 2022, p. 123). The *Holothuria scabra* attracts a varied multispecies community:

> Adults of *H. scabra* provide shelter to at least 17 species belonging to 4 metazoan taxa, i.e. annelids, crustaceans, gastropods and fishes, which live on their external surface or inside their coeloms and respiratory trees (Table 13). Most of the species associated with *H. scabra* were not found to adversely affect their hosts.
> (Hamel et al., 2022, p. 123)

Pearlfish is known to live inside the anus of *H. scabra,* which is known as a toxic sea cucumber, due to its secretion of saponins (Hamel et al., 2022, p. 120).[2] The pearlfish enters through the anus and lives in the sea cucumber's coelomic cavity or respiratory tree, having adapted itself to this symbiotic relation through an elongated shape and mucus that makes it resistant to the host's toxic secretions (Hamel et al. 2022, p. 128).

In symbiotic multispecies communities, communication is central: "communication between the associates is needed to ensure the maintenance of the symbiotic relationship through time" (Hamel et al., 2022, p. 130). It is especially through the chemical sensing of secretion in the seawater that these different species interact, and this is also how the *H. scabra* communicates with one another. While the secretion may be toxic to some species, thus keeping predators away, it can attract and strengthen others.

We can now appreciate Ingold's insistence on humans and animals to be "*differently* intelligent" (Lestel, 2002, cited in Ingold, 2022, p. 306). He elaborates on animal intelligence:

> And its intelligence is not an interior cognitive capacity, of which its actions are the effects, but lies in its whole way of perceiving and acting in the world. Each animal is different, but these differences are constituted in and through its entanglement in the generative process of social life, they do not exist in spite of it.
>
> (Ingold, 2022, p. 306)

By now we should have reached well beyond the anthropocentric understanding of cognition and communication, for a deeper appraisal of the interdependent dynamics of relational becomings. The question is not whether sea cucumbers are intelligent, but rather how they are differently intelligent? Nor do we need to wonder if they are social creatures, but rather recognize how they are social, with other sea cucumbers as well as other creatures in their lived environment.

Sensory Worldings and Sentient Becomings

Since sea cucumbers live by their senses, we can appreciate their engagements with their environment as a form of sensory worlding. Sea cucumbers know their life-worlds and interact with other ocean beings by way of their senses. Their sensory knowledge enables them to perceive and process their sustenance, protect themselves from danger and move around in their seawater environment. Their sensory worlding encompasses sensory forms of communication as well, with other creatures. It would be rather preposterous not to recognize the sensual intelligence that goes into their sensory worlding.

Can their sensory worlding help sea cucumbers become valued as sentient beings? Possibly, it depends on who you ask would be the short answer. Because sentience is a quality that is in the eye of the human beholder. In her seminal study *Becoming Salmon*, Lien (2015) explores how salmon become sentient beings as a result of human engagement through aquaculture. She approaches sentience as a "relational quality", rather than a property of salmon, and discusses how it is enacted in biological research as well as discourses on animal rights and animal welfare (Lien, 2015, p. 127). Lien shows how sentience is a fluid category, which tends to be "species-specific", focusing on whether specific types of animals can feel pain and suffer (Lien, 2015, p. 130). She notes that advances in the study of animal cognition, including fish, have expanded the notion of sentience to include relational practices in intraspecies interaction. If salmon are hard to assess as sentient beings by virtue of being fish, what about sea cucumbers?

A few years ago, the government of the United Kingdom recognized crabs, lobsters and octopus as sentient beings, which resulted in an amendment to the Animal Welfare Bill.[3] This recognition was a result of scientific research findings from the London School of Economics and Political Science (LSE). Through laboratory observations, a better understanding of the "intricate behaviors of specific invertebrates, particularly octopuses (Cephalopods)", helped scientists to identify "individuality and observable behaviors (personality traits) in these creatures", which in turn prompted their incorporation into ethical discussions (Crespi-Abril & Rubilar, 2023, p. 2). They underlined how the protection of octopuses in science use was legislated by the EU in 2013, thus preceding the UK legislation, which extends beyond the use of octopuses in research.

More recently, scholars have started discussing the sentience of echinoderms, including holothurians/sea cucumbers (Crespi-Abril & Rubilar, 2023). While recognizing the challenge of determining whether these invertebrates are sentient, they argue that "a lack of understanding of invertebrate behaviors does not preclude their capacity for sentience or their ability to respond to negative experiences in a non-anthropocentric manner that could cause suffering" (Crespi-Abril & Rubilar, 2023, p. 2). Although this discussion focuses on the use of sea cucumbers and other echinoderms in science research, it is an important step toward a more ethical appraisal of these creatures and by extension, greater protection of their welfare. As shown in the example of the octopus, animal welfare legislation that starts in science research can later be extended to other settings.

While the classification of animals as sentient beings builds on Western science, we can compare it with how animals are positioned in *African epistemology*. African ontology is founded on a *hierarchy of beings*, with God in the apex, then spirits and ancestors, followed by living humans (ranked by seniority), animals, vegetables and minerals (Chemhuru, 2023, pp. 94–95). Focusing on environmental ethics, some scholars have interpreted this hierarchy of beings as anthropocentric, and therefore incommensurate with animal rights (Horsthemke, 2017). But other scholars insist that this hierarchical order does not deny the value of animals, it actually constitutes a biocentric or eco-centric, rather than anthropocentric, ontology (Etieyibo, 2017). Meanwhile, emphasizing the *relationality* of African ontology, others have suggested that philosophies such as Ubuntu transcend the anthropocentric and eco-centric binary (Le Grange, 2019). Indeed, scholars have insisted that being African is always *being-with-others*, since the self is "substantially relational, largely constituted by interaction with other persons (and the environment)" (Metz, 2023, p. 70). Metz underlines how this Afro-relational ontology differs from the Western-individualist ontology in fundamental ways, emphasizing the interdependence of all life. This relational ontology can be extended to nonhuman beings as well, since they form part of the environment that humans interact with, even if they are considered to be below humans in a hierarchical order.

African relational ontologies exemplify the interconnectedness of human and non-human life, with a pronounced spiritual dimension. As depicted in the hierarchy of beings, spirits form an important part of life-worlds. Since the spiritual realm is part of African ontological realities (Chemhuru, 2023), it is also integral to the environment, which is both material and spiritual (Uimonen & Masimbi, 2021). In African cultural contexts, what some would refer to as *environmental spirituality* (Eisenstein, 2019) is thus part and parcel of life itself, an ontological reality far beyond a system of belief (Uimonen, 2023). This also means that the notion of sentience can encompass all living beings.

If animals, plants, soil, water, mountains, rivers, and so on are *sentient subjects*, we can no longer in good conscience treat them as instruments of human utility. We must take into account the well-being, integrity and even the dignity, of all beings, and not treat them as mere 'resources'.

(Eisenstein, 2019, p. 158, emphasis added)

I have elaborated on different appraisals of animals as sentient beings to broaden our understanding of the sensory worldings of sea cucumbering. Following Lien, we can appreciate sentience as a relational quality that is enacted, or not, from different cultural perspectives. Building on Ingold, we can also appreciate that all animal beings are relational, and they exist with other beings in different hybrid communities. So, sea cucumbers' sensory worldings in seawater environments constitute yet another instance of becoming-with, as Haraway would have it. This is how the sea cucumber lives its form of life, sea cucumbering in the ocean, making sense of its world through its senses.

The question of sentience is intertwined with the political ontology of human-animal relations. In his elaborations on pluriversal politics, Escobar notes that "most worlds live under ontological occupation" (Escobar, 2020, p. xxxi). He argues that ontological occupation is performed through the "categories and hierarchical classifications historically deployed by governments, corporations, organized religions, and the academy". The classification of certain animals as sentient beings can be interpreted as an example of ontological occupation by a modernist classificatory regime that uses scientific claims to differentiate the value of animals. After all, only some animals are awarded this status, most are left out, which is quite different from a more relational ontology that recognizes the sentience of all living beings. Escobar promotes the radical relationality found in other forms of ontological politics, for instance, Ubuntu, as civilizational alternatives to the modernist ontology that has led to our current environmental disaster (Escobar, 2020, p. 32).

Unfortunately, the ability to enact relational ontologies is curbed by dominant power structures, which privilege a materialist and utilitarian appraisal of sea cucumbers, recasting them as marine resources rather than as sentient ocean beings. As much as people in Tanzania who engage with them understand and respect the needs and preferences of sea cucumbers, they are not immune to the dictates and allures of global capitalism. Similarly, as much as scholars invest their time and effort into creating and sharing knowledge about sea cucumber morphology and behavior, their work is not isolated from broader societal structures, including market forces.

Sea cucumbers are thus caught up in contradictory ontological claims, curtailing their space for living their lives, sea cucumbering on the seafloor. Even so, they try their best to attend to their environment, as they are meant to. These brainless creatures are obviously not senseless.

Ironically, it is animals endowed with brains, humans who suffer from a superiority complex, who are messing up the ocean that sea cucumbers try to take care of. So what use is a brain if you have no sense?

A Love Story Beneath the Waves

The tide was gentle, and the seabed bathed in the soft glow of a full moon filtering through the water. For days, subtle changes had been brewing—shifts in temperature, currents, and light—whispering a promise of something extraordinary. For the two sandfish sea cucumbers, Holothuria scabra, this night was destiny.

Meeting Amidst the Sand

She was nestled in a patch of seagrass, her leathery body blending seamlessly with the sandy floor. He had been grazing nearby, his slow, deliberate movements leaving trails of cleaned sediment in his wake. Though their kind rarely interacted, tonight was different. The chemical signals released into the water spoke of shared purpose, and he found himself drawn to her presence.

A Silent Synchrony

They didn't touch; they didn't need to. Their connection wasn't one of physical proximity but of synchronization, a deep biological rhythm orchestrated by the ocean itself. Rising slightly off the seabed, they each extended their tentacle-like structures, sensing the water around them. The currents carried their signals, a delicate exchange of pheromones that said, "I am ready."

The Release

At the peak of the tide, the moment arrived. Together, they released their gametes—eggs from her, sperm from him—into the open water. The currents became their ally, carrying their genetic material away in a swirling dance, blending and mingling in the moonlit sea. Their contribution was now part of the vast ocean, a tapestry of potential life suspended in the gentle embrace of the tide.

The Aftermath

As the waters settled, the two sandfish remained still, as though reflecting on what had just transpired. The bond they had shared, ephemeral and instinctual, lingered like the soft caress of the current. They would part ways, returning to their solitary lives, but their union had left a mark—a new generation of sandfish drifting as plankton, unseen yet full of promise.

A Love Without Words

Theirs was a love born of nature's rhythms, a story of connection that didn't need grand gestures or spoken words. It was a love carried by the currents, etched into the water's flow, and written in the sand—a quiet yet profound testament to the enduring cycles of life beneath the waves.

(https://chatgpt.com/)[4]

Notes

1 ChatGPT prompt: Explain why *Holothuria scabra* buries itself into the sand, from its own perspective. 27 November 2024.
2 It was Rasmus Rodineliussen, a fellow ocean anthropologist (2024) who drew my attention to how pearlfish live in sea cucumbers. This odd symbiotic relation is also depicted in videos on YouTube, for instance productions by BBC, see https://www.youtube.com/watch?v=K2Eyup8Jk3w.
3 https://www.gov.uk/government/news/lobsters-octopus-and-crabs-recognised-as-sentient-beings, published on 19 November 2021, accessed on 26 November 2024.
4 ChatGPT prompt: Describe a love story between two sandfish during reproduction. 29 November 2024.

3 Suffering Domestication
How Would You Feel About Being Trapped?

Farming Holothuria Scabra in Ocean Pens

Holothuria scabra, commonly known as sandfish, is a species of sea cucumber highly valued for its role in aquaculture. When farmed in ocean pens, its experience depends largely on the conditions provided by the farmers, including pen design, stocking density, environmental factors, and management practices. Here's a breakdown:

Behavior and Habitat Suitability

- *Natural Habitat Mimicry*: sandfish are detritivores, meaning they feed on organic matter in sediment. Successful farming in ocean pens involves replicating their natural seabed habitat with appropriate sediment quality and depth. If the conditions are similar to their natural environment, they are likely to thrive.
- *Space and Movement*: sandfish require sufficient space to forage and move. Overcrowding in pens can lead to competition for resources, stress, and increased susceptibility to disease.

Environmental Stressors

- *Water Quality*: good water flow is critical for maintaining oxygen levels and removing waste. Poor water quality, often due to stagnant water or pollution, can cause stress and health issues in sandfish.
- *Temperature and Salinity*: sandfish are sensitive to extreme temperatures and salinity fluctuations. Rapid changes in these parameters, common in poorly managed pens, can lead to physiological stress.
- *Predation*: although pens can protect sandfish from some predators, if not well-constructed, they may still be vulnerable to predation by crabs, fish or birds, leading to stress and mortality.

Human Interaction

- *Handling Stress: frequent handling, such as during harvesting or pen maintenance, can cause stress or physical damage to sandfish, which are soft-bodied animals.*
- *Feeding Practices: while sandfish are self-sufficient in nutrient-rich sediments, additional feeding in some farming systems can alter their behavior and health. Proper feeding practices can enhance growth but must avoid overfeeding, which degrades sediment quality.*

Health and Disease

- *Disease Risks: farming in high densities can increase the spread of diseases, which may not occur as commonly in the wild. This is a significant concern for sandfish farming.*
- *Sediment Degradation: accumulation of waste in ocean pens can alter sediment quality, impacting the health and growth of sandfish.*

Adaptation and Growth

- *Sandfish tend to adapt well to ocean pens when environmental and nutritional conditions are managed effectively. In well-maintained pens, they exhibit normal growth rates and reproduction, suggesting they are not overly stressed.*

Ethical Considerations

- *There is growing interest in understanding the welfare of farmed invertebrates like sandfish. While they lack a central nervous system and the capacity to experience pain as vertebrates do, they do respond to adverse conditions through behavioral and physiological changes, indicating discomfort or stress.*

> *In summary, Holothuria scabra can experience farming positively if conditions mimic their natural environment and stressors are minimized. Conversely, poor management can lead to stress, poor health and reduced growth, underscoring the importance of adopting best practices in sandfish aquaculture.*
>
> *(https://chatgpt.com/)[1]*

I find it rather impressive that AI lists the importance of good management when sea cucumbers are farmed in ocean pens, thus highlighting the interdependence of humans and non-humans in aquaculture. It ties in well with anthropological research on domestication that decenters dominant narratives of human mastery over nature, bringing forth the interdependent dynamics of human-animal relations, especially the significance of non-human agency (Swanson et al., 2018). To capture these more-than-human entanglements,

Lien (2015) appreciates domestication in terms of mutuality, uncertainty and tinkering. She notes that aquaculture is a particularly interesting case of domestication since animal husbandry practices are extended to marine life forms – in her case, fish – which can be thought of as "newcomers to the farm" (Lien, 2015, p. 2). There are some interesting similarities as well as differences between salmon and sea cucumber aquaculture.

Figure 3.1 (a–c) Sea cucumber farms of varying sizes with different organizational set up. Photographs by author.

This chapter explores sea cucumber farming, through ethnographic engagements with farms in different locations in Tanzania: on the mainland and in the Zanzibar archipelago. It shows how both sea cucumbers and humans are transformed in the process of domestication, the former becoming kept animals that require human care, business and livelihood for poverty alleviation, while the latter become ocean farmers and blue entrepreneurs. Detailed ethnographic descriptions of sea cucumber farms unravel the complexities of these multispecies worlds in their multimaterial formations, linking land and sea in intricate ways. The challenges of reproduction are discussed at length, from the capture of fingerlings in the wild to science-based reproduction in a hatchery. Some ecological risks of aquaculture are laid out, highlighting the uncertainties involved, especially in relation to environmental changes brought about by human intervention. The chapter concludes with an AI-generated poem on the impact of climate change from a sea cucumber's perspective.

Sea Cucumber and Human Becomings in Artisanal Aquaculture

In her seminal study of salmon aquaculture, Lien tells stories of *salmon becomings*. She recognizes that "the story of salmon can be told as industrial success or as environmental catastrophe", and that both versions are true, yet partial narratives, so she adds her own "story of salmon as a story of becomings" (Lien, 2015, pp. 166–167). Lien approaches domestication as an ongoing process of relational practices, showing how aquaculture enacts biosocial formations through various becomings of salmon: as hungry, biomass, scalable, sentient and alien.

Comparatively, through aquaculture, sea cucumbers become *kept livestock*, similarly to how the keeping of other kinds of animals entails various practices of domestication. In Swahili, the word used for farming sea cucumbers is kufuga (to keep/raise livestock), which is the same verb used for keeping domestic animals (typically chickens, goats, cows). In this predominantly agricultural society, keeping animals is quite common. Even semi-urban households will raise some chicken, which often walk freely and peck the ground around the house, and beyond. Keeping sea cucumbers is a new practice, which has been introduced by the government and other development actors, as a way of safeguarding dwindling amounts of wild stock, while providing alternative income opportunities to coastal communities.

When sea cucumbers become *kept animals*, they require practices of *care*. This care work requires paying close attention to the needs and preferences of sea cucumber, thus engaging in care as *maintenance work* in multispecies worlds (Puig de la Bellacasa, 2017; Uimonen, 2025). When caring for animals, human caretakers tend to rely on *experiential knowledge* (Singleton, 2010), through *embodied practices* of attentiveness (Mol et al., 2010).

> When we asked them about *jongoo* as a creature, Rashid said it was to farm, like *kuku* [chicken]. You had to take care of them so you could benefit from them. They also need their rights, so farm with care, like to clean the farm, so they can get their food in a good way.
>
> (Fieldnotes, Pemba, 23 July 2023)

Rashid was Chair of a small group in Pemba that built a sea cucumber farm in January 2023. The group was established on 31 October 2022 (he checked exact dates on his smartphone during our interview), with seven members (two women and five men). The farm was built with TZS 2 million from one of the members, then TZS 1,350,000 was added to buy fingerlings. The farm measured 40 × 40 meters, and they started with 1,160 fingerlings. They got the idea for the farm from a fellow member, who said sea cucumbers fetched a good price. Rashid told us that in Pemba sea cucumbers were bought for TZS 60,000/kilo (about USD 24), but on the world market, it fetched a high price, up to USD 1,000 for a kilo in China, as he had seen on a film from Madagascar. He found the video on his phone and showed us. He had heard you could get TZS 300,000/kilo in Tanzania, even up to 700,000, if dried. Their aim was to sell dried sea cucumbers, and they showed us some they had already processed, dried for several days. Unlike most sea cucumber farmers, Rashid was quite well educated with a college certificate in tourism, and he spoke some English. When we first met him, he was 40 years old and married, with eight children. His profession was fishing and farming. During our interview in his house, some chicken walked in through the door, pecking on the floor, undisturbed.

Sea cucumbers also become *business*, since the primary motivation for keeping them is to profit financially, rather than to produce food for local consumption:

> "Jongoo? Biashara nzuri" (Sea cucumber? It is good business)

This is how Zuwena, a woman member of the sea cucumber association in Kaole replied to our question on what she thought about the sea cucumber, during our interview in January 2023. We had been talking about the sea cucumber farm, so it was no wonder that she immediately thought of the jongoo as business. It was only after we prodded her about the jongoo as a creature that she elaborated on it as a creature created by God (see chapter two). Zuwena was a widow in her late 50s, and she lived with four grandchildren. She had completed primary education at the local school in Kaole. Zuwena sustained herself and her household with a small business of cooking samosas and porridge, for a daily profit of TZS 2,000 (USD 0.85). She also earned about 20,000–25,000 (around USD 10) per week through her tondo (sea snail) business. Like some other women in Kaole, she picked tondo in the mangroves, processed and sold it in nearby towns. She also used it for food. In poverty-ridden coastal communities, sea cucumbers become a means of *securing livelihoods*, to *alleviate poverty*:

> Based on the knowledge they get in training, the *jongoo* came to save their lives economically, alleviating poverty, Abuu responded, when we asked him about *jongoo* as a creature. If government has a good plan, starting from the grass roots, this is good, this blue economy, he reflected. If we plan and the government can organize, this can work well and make life better for coastal people.
>
> (Fieldnotes, Mtwara, 16 November 2023)

Abuu's reflection was interesting, as it conveyed what the sea cucumber became during training for farming. We were sitting at the FETA offices in Mtwara. Abuu had approached us on the bench earlier, with a bucket with some jongoo of different sizes. He had picked them outside the farm, but was going to put them inside it to grow. We asked him if we could interview him, and he agreed. He was part of a group of 18 people (nine women and nine men), now being registered as an association. Some members had been trained by FETA, and they had started their sea cucumber farm in July 2022. The farm was 30 x 60 meters, with 1,200 fingerlings, planted in October 2022. Abuu was 58 years old, married, with five children.

These examples illustrate how *artisanal aquaculture* enacts sea cucumbers in different ways, but what about the *becomings of humans* as sea cucumber farmers? From a multispecies relational lens, we need to consider how humans are enacted through aquaculture. After all, if domestication is "a set of dynamic and mutual multispecies relations", it is not enough to "include nonhumans in the story of the Other" (Lien, 2015, pp. 164, 168). Lien (2015) has discussed this mutuality in terms of how salmon become domesticated in the care of farm workers and how they in turn become domesticated by the salmon (p. 47). In Norway, salmon aquaculture has developed from "small-scale experiments by local entrepreneurs", to a state-supported salmon industry with a highly regulated work environment for farm workers (Lien, 2015, p. 46).

In Tanzania, through artisanal aquaculture, *fishers become ocean farmers*, which entails a relocation and reformulation of farming practices. Coastal communities have always combined fishing and farming, drawing their livelihoods from the sea as well as the land. Farming has often been small scale, on privately or family owned farmland in the vicinity of the village. It has also entailed animal keeping, typically chickens, goats and cows. In Pemba, farmland was often located next to the ocean, in flatlands below villages. In mainland Tanzania, villages are often near the ocean, with farmland further inland.

Keeping sea cucumbers means that farming takes place in the ocean instead of on land, since aquaculture constitutes marine rather than terrestrial farming practices. It also entails keeping animals that are not fully domesticated, their confinement in ocean pens being but the first step in the process of domestication. In this sense, we can think of artisanal aquaculture of sea cucumbers as *partial domestication*. Unlike terrestrial animal farming, keeping sea cucumbers entails various caring practices related to their particular needs, as discussed above. It also means that their well-being depends on oceanic conditions which are well beyond human control.

In the context of the blue economy, sea cucumber farmers are expected to take care of sea cucumbers, keeping them well, as Rashid reflected. But the motivation for this care is above all financial, a good business, which is how Zuwena enacted the sea cucumber. Unlike other seafood collected or fished in the ocean, the sea cucumber is farmed for export, with China as the main destination market. Since this business is framed within the blue economy model of capitalist exploitation and extraction of ocean resources, the sea cucumber becomes a commodity in the global seafood trade. In the global blue economy, artisanal sea cucumber farmers in Tanzania are positioned at the lowest rungs of global power hierarchies, serving as vectors for blue entrepreneurship in the peripheries of global capitalism. Here, the business of sea cucumbers is about livelihood, about poverty alleviation, as Abuu reflected. Even so, as entrepreneurs, they have little financial security, while the risks involved are considerable, for humans as well as sea cucumbers.

Multispecies and Multimaterial Ocean Farms

Spatial confinement is a fundamental aspect of aquaculture, ensuring that farmed ocean creatures stay in place, under human supervision and control. Indeed, it is the separation of wild animals from farmed ones that is a defining feature of domestication, which in turn rests on ideas of the separation of nature/culture, human/nonhuman and so on (Swanson et al., 2018). This separation holds true for aquaculture as well: "fish domestication could be defined as a dynamic and endless process, which starts as soon as individuals are transferred from wild to captive conditions" (Teletchea, 2021, p. 87). These captive conditions vary, depending on species, location, resources, technologies and other factors. As detailed in Lien's study, industrial salmon aquaculture in Norway rests on and reproduces this nature/culture divide. As for artisanal aquaculture of sea cucumbers in Tanzania, these coast-near practices tend to reproduce the interconnectedness of land and sea, which is a defining characteristic of this amphibious socio-ecological context (Uimonen, 2025).

In industrial aquaculture of salmon in Norway, spatial confinement is elaborate and high-tech. Lien discusses the caged home of salmon in terms of *domus*, a heterogenous multispecies assemblage. The material structure of salmon farms is "robust" and "intricate", like in Vidarøy, where "some 500,000 salmon are being held in place by nets attached to an arrangement of rollers and cables, hooks, and eyes" (Lien & Law, 2011, p. 81). This industrial set up includes a house with various machinery, while the salmon are kept in "ten cages (and one pen) connected by metal walkways and a platform" (Lien, 2015, p. 49). The farm is located in water, but it is only accessible by boat, and the people who manage the farm do so from platforms above the surface. Lien (2015) discusses the material structure in Vidarøy as a "complex material interface" that also defines and mediates various "boundaries", as exemplified by the netting of cages that limit the movements of salmon (p. 57). But far from constituting a clear-cut boundary, the netting is porous, keeping larger fish out, while letting smaller ones in, along with water as well as marine parasites. Algae may also grow on the netting, which reduces its permeability. This is dealt with by rolling up the netting from the water to dry, which is part of "regular maintenance work" (Lien, 2015, p. 58). The netting also erects a boundary between farmed salmon and wild salmon, but these boundary practices aimed at holding nature and culture apart are laborious and uncertain (Lien & Law, 2011).

Sea cucumber farms in Tanzania are both similar to and different from Norwegian farms – their material structure being low tech and practices more hands-on. There is considerable diversity in size and organizational set-up, from small collectively run farms of a few hundred square meters to massive privately owned farms covering thousands of square meters. Sea cucumbers do not need to be fed, but humans have to make sure there is enough food in the ocean pens where they are kept. These pens are bordered with a fence made of galvanized iron pipes and/or wooden poles (mangrove) and green plastic mesh. The fence runs along the seafloor, which is where the sea cucumbers live, feeding on the sandy-muddy sea bottom. The material height of the fence varies, as does the tidal water level, but the pen is mostly underwater. The borders do not always contain the sea cucumbers in their captive space, as some escape. Nor do they keep other creatures out, such as fish, which in turn also attracts fishers, who are not supposed to access the area. The fences also capture things that do not belong, such as leaves blown by the wind. Algae grow on the plastic mesh, and strong winds can overturn the fences or poles, which are also worn down by the waves and saltwater.

Maintenance is done through manual labor, as farmers enter the water during low tide to clean and clear the pens, using muscle power and simple tools like machetes, ropes and brushes. Both men and women engage in such tasks. Sometimes men with diving skills assist in the maintenance, free-diving with mask, snorkel and fins around the pens.

Security patrol is a recurring task, one of the most important ones, to make sure that other humans are kept away from the ocean pens. Patrolling is done on foot, by boat or from watchtowers; in larger commercial farms, more advanced equipment may be used, such as CCTV cameras. It is a complicated task, since sea cucumber farmers have to watch out for members of their communities, often neighbors or relatives, who may be trespassing into the farm area, whether knowingly or not. There is also a considerable risk of theft, since sea cucumbers can easily be sold for some cash through local traders.

We walk to the sea cucumber farm in Kaole, barefoot on the seafloor, which is exposed during the low tide. The leisurely walk only takes some 10 minutes, on the soft sea bottom of sand and mud, with puddles of water. It is 10 November 2022, and I have just returned to Tanzania for follow-up fieldwork. In preparation of our visit, we have talked with Abraham from the sea cucumber association, who told us they had been picking fingerlings for the last few nights – except yesterday, because there was a big football match. We had suspected something like that, having seen a lot of flickering lights at night, people with torches in what seemed to be the farm area. Three ladies are on duty when we reach the farm. We walk around the pens with Abraham.

Abraham was picking up some jongoo, small ones that he shifted to the pen where they keep fingerlings. The fingerlings were 6-8 centimeters long. He was also cleaning around the fence a bit. The water was a bit murky and very low. It was warm, but not as warm as near the shore. It was very sunny. We could see some new mangrove sticks used to support the fence. Older sticks had small shells growing on them. There were also leaves floating on the surface, which gathered in the corner of the fence. Many jongoo were hiding under the sand, we could sometimes see a bump on the bottom indicating a jongoo. Abraham dug some up, showing us.

We asked how many species of jongoo there are, Abraham said only one species is used for food. Others are not for business, but are important for the ecosystem. After a while, he came over with a jongoo that had hard bumps on its skin, it's called "jongoo tairi" he explained, meaning tire jongoo. He returned it where he found it, saying it was important for the other jongoo. We also saw some white, long, snake-like things that looked a bit like jongoo, but they were worms and left alone.

(Fieldnotes, 10 November 2022)

Sea cucumber farms are *multispecies worlds* that are becoming with a diversity of seawater life forms, animals and plants. The main species is the sea cucumber itself, more precisely the *Holothuria scabra* (aka sandfish). But other sea cucumbers may also venture into the farm, like the tire species

Figure 3.2 (a–c) Manual maintenance work. Security through CCTV camera or watch tower. Photographs by author.

Abraham picked up. Apart from sea cucumbers, I have also observed small crabs, fish and jellyfish, as well as sea worms and sea stars. Once I saw lionfish around the fences, some inside the pen, slowly swimming around, their spectacular fins spread out. Inside the pens, seagrass grows in patches on the sandy and muddy seafloor.

These multispecies worlds are also *multimaterial becomings*, composed of organic and non-organic material. The manmade structure of the farm is created out of plastic, wood and iron, a semi-robust infrastructure. Humans have brought these alien materials to the seascape, to erect boundaries around the pens where sea cucumbers are kept. The fence is a green mesh of solid plastic, with large holes to let water flow through. It is fastened to poles with thin plastic string or loops. The fence is stretched across the sea bottom, reaching upwards, with a height from half a meter to a couple of meters, depending on the water levels. The poles are made of galvanized iron or mangroves, which have been cut to size.

Multimaterial ocean farms create new habitats for various species. Different varieties of algae often grow on the plastic fences, eventually weighing them down, while obstructing the flow of water. Barnacles clutter along the iron and mangrove poles that hold the fences. Squid sometimes lay eggs in the fences or the poles, thus adapting these foreign materials to their nesting habits. The fences also capture things floating on the ocean, not least natural debris such as fallen leaves, which tend to gather in the corners.

It is as if the ocean claims terrestrial materials, adapting them to the marine environment, rather than the other way around. As much as humans try to control the ocean space, their tools do not seem to stand much of a chance against the dynamic vibrancy of seawater. The ocean claims and reclaims its space, intertwining manmade materials from land with living marine organisms, who find new habitats in these alien constellations. Plastic mesh to erect borders around farm pens? Algae grow, take over, and redirect water flows. Iron poles to mark and solidify manmade boundaries? Barnacles like to live on manmade objects, iron poles now offering new habitats. Mangrove poles removed from the intertidal land-sea borderland to keep sea cucumbers at bay in the ocean? Wooden structures worn down by water and wind, now home to clusters of sharp barnacles.

Even so, the spatiotemporal rearrangement of ocean space testifies to the human ability to exert power over seascapes, while limiting the agency of ocean creatures in their seawater environment. As much as the ocean reclaims its space, its agency is limited. The manmade borders of the ocean farm affects the mobility of marine life forms, especially sea cucumbers. Entrapped by manmade borders, they cannot move freely on the sea bottom, especially when they need to avoid danger. As we will see, this limitation can have disastrous effects for sea cucumbers. By extension, it will also affect the ecosystem they are devoted to taking care of.

Suffering Domestication 75

Figure 3.3 (a–h) Multispecies sea cucumber farms. Photographs by author.

Wild and Farmed Fingerlings for Reproduction

Reproduction is a critical aspect of aquaculture; some even see it as a defining characteristic of domestication. While recognizing that a domesticated fish is "neither a definitive status […] nor a final end point of domestication", some scholars have argued that "Domesticating a fish species implies that the full life cycle is controlled in captivity without wild inputs" (Teletchea, 2021, p. 88). By comparison, although the cultivation of salmon had been tried out for some 150 years, it was only after experimentation with floating cages in the 1950s that the first batch of salmon completed an entire life cycle in captivity (Lien, 2015, p. 34). In Tanzania, the partial domestication of sea cucumbers is best appreciated as an ongoing, incomplete and uncertain process, since full life cycles in captivity is difficult to reach. So far, most sea cucumbers have been grown from fingerlings collected in the wild or in farms.

When we walked to the farm that morning, Abraham told us about a large order for fingerlings from Mtwara, which was putting a lot of pressure on the Kaole association. Kaole had harvested in September, so it was time to replenish the stock. Meanwhile, Mtwara was just starting sea cucumber farms and a government representative had promised them that they could get the initial fingerlings from Kaole, which was supposed to have plenty. Since it was initiated by high level government officials, even a Minister was involved, this deal put great pressure on the Kaole farm to deliver more than they were capable of.

Researchers: Is this not the time for having fingerlings?

Abraham: From December to January there will be plenty. Those from Mtwara who need 400,000 fingerlings are those who make us tired. They were brought here by the government, so it's better you tell them there are too few, than telling them that there's nothing. Those people they already put TZS 14 million in our account, and they will continue to add until it reaches 100 million, because they need 400,000 fingerlings.

Researchers: So now you just collect here, then you see how to transport them?

Abraham: Yes, our target was to get 10,000 fingerlings by today. Today is Thursday, tomorrow Friday. We wanted to see how we transport, now the plane is expensive, so they said we take a lorry from Bagamoyo, the bags for carrying are already ordered somewhere.

Researchers: So now, how many are needed?

Abraham: At least 10,000, to appease them [since they want to appear as if they have more than they have].

Researchers: And now, how many have you caught?

Abraham:	Like 5,000 or so. Today is the 15th in the month [end of low tide], now the water will start to come back and it will not be leaving much, so now that water becomes *mafu* [up to the thigh] we will find an alternative way, to talk well with divers and see how it can be done [to reach the 5,000 missing].
Researchers:	So are you sure that in the first round they are going to have those 10,000?
Abraham:	10,000 they will get, because when they get it, because when you give them, you give them the good language [excusing language], like it didn't rain [blame the weather].

We fear to give them unexpected information because they came with the Minister. Ulega [Minister of Livestock and Fisheries] came with them, so we can't know what the future plan is.

(Interview with Abraham, Kaole, 9 November 2022)

The relations between the Kaole and Mtwara farms exemplify multiple stakeholders involved in sea cucumber farming – in this case, communities and government. The Kaole farm is managed by the local Beach Management Unit (BMU), which functions like a local authority, tasked with protecting the ocean environment. As such, the BMU is expected to follow and implement laws and regulations. But BMUs have little power, so their ability to act is limited. Since they occupy the lower rungs in the state hierarchy, they need to maintain good working relations with other state agencies. The higher the authority, the more respect and subservience is expected, to smoothen relations. Thus, the ingratiating behavior towards the Minister and his Ministry, even if it means withholding the truth. In the midst of it, the Mtwara farms were initiated by FETA, a government training institute for fisheries, which ranks higher than BMU.

In Tanzania, state involvement puts considerable pressures on sea cucumber farms operated by coastal communities, assuring that farmers enact the market logic of the blue economy, while respecting government hierarchies. It is not uncommon that politicians make unrealistic promises, as exemplified by the Minister promising 400,000 fingerlings to Mtwara. But rather than challenging such powerful state actors, people lower down in the hierarchy will try different means of circumscribing these impracticable pledges. In this case, delivering a small amount to appease expectations, while delaying the disappointing reality of being unable to deliver what has been promised.

Meanwhile, sea cucumbers are dispossessed of their agency, subject to human care or negligence. While farmers negotiate challenges of acquisition and distribution, sea cucumbers are unable to get on with their sea cucumbering. Tragically, the 10,000 fingerlings that the Kaole association managed to collect died during transportation. Stored in a black plastic container on the back of a truck, they baked to death under the hot sun (Uimonen, 2025).

Marine Hatchery for Industrial Production

When the government started to encourage sea cucumber farming, the idea was to develop and replenish stocks through a marine hatchery in Zanzibar. The hatchery was developed with support from Korea (KOICA) and FAO, and inaugurated in April 2018. It did not operate well for several years and the supply of fingerlings to farmers was very limited. In January 2024, the hatchery was reorganized and is now operated as a public-private partnership between the revolutionary government of Zanzibar and a local company. The Zanzibari government also set up a large farm in Pemba, to be used as a nursery.

On our first visit in December 2022, the hatchery seemed semi-idle. The manager who showed us around explained the process, but there was little sign of activity, apart from a few small tanks holding a few dozen sea cucumbers. At the time, the focus was on tilapia fish, kept in large tanks. They were trying to breed a new kind that could live in seawater. They had managed to get it to survive and grow, but not yet to reproduce. The idea was to develop tilapia aquaculture in coastal Zanzibar, thus expanding it from freshwater aquaculture on the mainland to seawater aquaculture (mariculture) in the archipelago.

Figure 3.4 Entrance to Zanzibar Marine Hatchery. Photograph by author.

When we revisited the hatchery in May 2024, it had developed quite a bit, with improved infrastructure and active production of juvenile sea cucumbers.

The hatchery was visibly different, with a new sign at the entrance. The sign read *Zanzibar Marine Hatchery Company Limited*, with a government logo and a company logo. From the outside I could see that they had covered the walls with some sheets, when I got closer I could see that they were made of plastic, but they looked like corrugated iron sheets.

The Director was a bit hesitant about letting us in, but after a while he allowed us to enter and film, with the guidance of his staff. There were so many of us, he feared contamination, he said. They had just gotten 22 million eggs from spawning, up from a few days ago, when they had reached 17 million. But survival rate was only 3%.

When entering the main building we had to step into a small patch of water, to cleanse our shoe soles. I remembered it from last time. But now that the walls were covered (instead of half open), it made more sense. When we entered, it smelled glue. Some men were building tanks with fibre glass, shaping and fastening thin white sheets with glue, then pasting them with brushes. The smell was strong.

On our right were long tanks, with water going into them in a thick pipe, and washed out at the bottom of the short side. The man adjusted and switched off the water. Inside the tanks there were sea cucumbers of varying sizes. The water was aerated, so the surface was not calm, but still they were visible. Some were buried into the sand, others just lay around, sometimes in clusters.

After a while I could see one of them moving, pressing its body upwards for each move, and slowly lurking forward. It looked like it had some tiny feet on the bottom. I was fascinated to see it move and filmed it. Later on, I got even better footage of another one that was moving, more quickly.

Here in the tanks with clear water it was actually easier to observe the sea cucumbers in their seawater habitat. The water was translucent and the bottoms of the tanks were lined with sand. In one tank I noticed some small sandy balls lying in strings, and yes, it was their poop. It was a good sign, showed they were eating, the man explained. Hussein had watched one move and said it had also pooped, then continued moving. He was impressed by the speed. Within minutes it moved several inches.

We were taken into a closed room that was very warm and humid. The glass door and windows had water running down from the humidity. I immediately felt the warm dampness, like a sauna. In the round tanks were "babies". What looked like specks on the bottom were 2 months old, in

another tank tiny juveniles about 1-1.5 cm long, they were 3.5 months. It would take them about 5 months to become fingerlings, depending on the conditions, a man explained to us, in English. He wore a t-shirt with LOMBOK on it and had straighter hair and lighter skin. I asked where he came from? Madagascar!
Outside the hot room was a whiteboard where they kept track of the stock.

(Fieldnotes, Unguja, Zanzibar, 20 May 2024)

The hatchery creates a *science-based approach* to sea cucumber aquaculture, enacting reproduction in a modern, controlled environment. Here the material infrastructure is composed of modern materials: customized fiber glass tanks, plastic hoses and pipes and hanging electric lights. The walls are built from concrete, covered by corrugated iron roofs with light inlets. The concrete floor has drains, covered with iron grids, allowing water to flow off the wet floor. There are few staff in the buildings, all of them look busy. They are all men, mostly young or middle-aged. Records are neatly kept, written on black boards and the sides of tanks, letters and numbers denoting activities and progress.

The scientific ambiance is enhanced by the closed rooms, the one for spawning described above, another one for feeding, marked Algae Room. In this room, algae for feeding are prepared and processed in a laboratory-like environment, complete with lab-like glass bottles. When reproduced in a laboratory setting, the larvae need to be fed in order to grow. From the outset, the food has been a special type of algae, imported from the United States. In the algae room, white metal racks hold glass bottles, with different colored liquids, closed with corks, with thin plastic pipes inserted through the top. The bottles are marked with marker pen, denoting date and type. Plastic buckets with green algae are also stored on the shelves. The room is insulated and the temperature kept high.

The hatchery represents the culmination of domestication, enacting reproduction under human control and mediated by technology, thus bypassing the need for wild stock. As you may recall, some scholars insist on complete human control over the reproduction process for fish to be considered domesticated, in other words reaching a stage where their "Full life cycle is controlled in captivity without the use of wild inputs" (Teletchea, 2021, p. 88). In the case of sea cucumber aquaculture, this stage requires the use of hatcheries, since it is the only way humans can ensure a controlled environment. In the open sea, the eggs will simply disperse far and wide, as Neema reflected on how the eggs move in the last chapter.

Being able to reproduce sea cucumbers in a hatchery setting is also crucial for sustainability, since the use of wild stock will not protect – and may even decrease – overall stocks. So far, most sea cucumbers grown in farms have been raised from fingerlings collected in the sea, only a fraction has been grown in a hatchery. This means that stocks have not really been protected. They have merely been shifted from the open sea to confined ocean farms. Meanwhile, the experience of those who have received some of their stocks from the hatchery have not necessarily been positive. In one farm in Pemba, fingerlings from the hatchery were combined with locally found ones. Unfortunately, most of the farmed sea cucumbers were stolen before harvest. Ominously, today there are few sea cucumbers left in the area. It seems as if the failed attempt at farming reduced their numbers overall.

Above all, the hatchery creates a processing facility for industrial reproduction, thus moving beyond and at the same time mediating artisanal aquaculture farms. If the hatchery were your only visual impression of sea cucumber farming in Tanzania, you would think it was undertaken with modern technologies at an industrial scale. But this is not the case, at least not yet. The hatchery is supposed to facilitate the reproduction of sea cucumbers for artisanal farms around the country, supplying stocks to the mainland as well as the archipelago.

But the vision is far grander, as the Director explained to us, a middle-aged man, well versed in the government. He had worked in the Forestry department, and was then promoted to environmental conservation, and after retirement to the hatchery. He was very optimistic and promotional in his style, keen on selling the idea of the hatchery and its prospects. When we interviewed the Director after being shown around the hatchery, he told us excitedly about their plans for the future.

The private-public-partnership (PPP) was clearly in it for business, a partnership between the government of Zanzibar and a company (AfriVision), run by Indians living in Arusha. The arrangement was 49% government and 51% private company. They had big plans. He had been recruited after his retirement to oversee everything, in January 2024. The hatchery had been busy since then, and they were still building it. If something is needed today, you must fix it today, he underlined, you can't wait.

When he spoke of future profits, he got the dreamy look in his face that I have seen in others, the LURE of BIG MONEY. He talked of 3,000 sea cucumbers sold at 10,000 each, after 8-10 months. You can do the math, he said [TZS 30,000,000, or USD 12,000]. If you come back in a month, you will already see changes, and in 6 months, even more so, he said. But "nothing is free". [He explained that fingerlings would be sold to communities, not given for free. This was obviously a business venture].

(Fieldnotes, Unguja, Zanzibar, 20 May 2024)

But marine hatcheries are also fraught with risks, indicating the limits of human mastery over marine life forms. Even if they manage to solve the problem of restocking, scholars have warned that hatcheries create new risks, from disease to genetic alteration, which affect both domesticated and wild stocks (Eriksson et al., 2012; Fabiani et al., 2023). Moreover, the eventual impact of domesticated sea cucumbers on the marine ecosystem is still unknown.

Risks and Uncertainties in Times of Climate Change

When we were driving to Mtwara, we received some bad news from Rashid in Pemba; they had lost all their 1,400 sea cucumbers in one day. He had texted Mary through WhatsApp and sent her photos of piles of dead sea cucumbers, their bodies shriveled up and flattened. The farmers were at a loss for what had happened. They guessed they had been poisoned. Mary suggested that maybe it was due to the heavy rains, too much fresh water. She remembered that the farm was not far from slopes that had erosion. We talked with FETA tutors in Mtwara about the incident, and they suggested the sea cucumbers had died from a disease. I sent the name and number of one of the tutors to Rashid, for consultation. He informed me that the government had also taken samples of dead sea cucumbers, but they had not yet received an answer. Months later, I mentioned the incident to an expert who had worked in Madagascar. He said the sea cucumbers had probably died because of environmental changes. It had rained heavily, so a lot of fresh water had entered the ocean, mixing with the sea water, thus reducing its salinity. The farm location was not good, he concluded. When we revisited Rashid's farm in May 2024, six months after the tragic incident, they had not yet received a satisfactory explanation from the authorities or experts. Meanwhile, they suspected that someone had poisoned their sea cucumbers. After all, why did none of the nearby farms have the same bad experience?

As you may recall, Rashid spoke well of the need to take care of sea cucumbers when farming them, even mentioning that they have their rights that need to be attended to. These caring practices included making sure there was enough food in the pen and keeping it clean, so the sea cucumbers could thrive in their seawater environment. Together with his partner, Mohammed, he had worked tirelessly to maintain the farm, for which they were compensated a small amount by their group, mostly paid in kind. When we first visited them in July 2023, Rashid and Mohammed had taken us to the farm by a borrowed fiber boat. Mohammed had agreed to dive and take pictures with our GoPro camera. He had also brought up a couple of sea cucumbers, so we could take pictures. They had been so proud of their work and so optimistic of their future success. But these unexpected deaths were beyond their care and control. Rashid was devastated by the loss of investment. When I checked up on him through WhatsApp in early December 2023, his response was rather desolate: "now is silence no one to help us we are tired lot of money disappeared we don't know how to do".

The suggestion by the FETA tutor that the sea cucumbers had died from a disease is instructive of some of the known risks in aquaculture, labeled as *biosecurity*. This tutor, Baraka, has an MSc in Marine Science from the Institute of Marine Science (IMS) at UDSM.

We showed pictures to the FETA team and Baraka immediately recognized a disease, a Skin Ulceration Disease (SKUD). He pointed to the white spots on the skin in the pics. It was caused by change in salinity, which caused a skin disease that attracted bacteria to the white spots, bacteria carried by mangrove snails.

This had happened in Madagascar in 2019.

That is why they teach *biosecurity*, how to prevent disease. They showed us on a certificate, which specified the different topics covered, including Biosecurity.

They explained some guidelines:

- There should be no mangrove snails around pens.
- And they need to maintain salinity. They need to check during several tidal periods before pen construction, to monitor any entry of fresh waters, for instance from rivers. There are instruments for this, used in salt farms.

(Fieldnotes, Mtwara, 13 November 2023)

The SKUD that Baraka mentioned is a disease spread by bacteria. It occurs in *H. scabra* aquaculture due to several factors, such as sudden changes in temperature, variations in salinity, abnormal drop in oxygen, the density of sea cucumbers or lesion left by predators such as crabs (Hamel et al., 2022, p. 227). In this particular case, Baraka attributed its spread to mangrove snails. All these issues can be related to the environmental impact of aquaculture, which confines sea cucumbers in enclosed spaces.

Lien has discussed how "domestication of salmon has microbial side effects", from proliferation of sea lice to "unknown viral and bacterial diseases". She concludes:

Disease outbreaks and parasite attacks are just two ways in which the salmon assemblage is fragile. Holding it together requires constant practices of knowing, caring, and tinkering, and mobilizing networks that extend far beyond the salmon farm. Salmon farming is precarious practice in which humans are active participants but are hardly in control.

(Lien, 2015, p. 74)

Biosecurity becomes an interesting way of conceptualizing and managing this lack of human control in aquaculture, while concealing the negative effects of human interventions in marine environments and the *aquabiopolitics* involved. Scholars have discussed how "biosecurity increasingly involves governing through uncertainty and insecurity", not least in relation to "unruly biological life" (Barker et al., 2013, p. 9). Concerns with biosecurity have risen in response to heightened mobilities from globalization, and more recently in relation to the unknown challenges of climate change. While biosecurity gives a semblance of human control, it also obscures how human interference causes environmental problems to start with. In this sense, biosecurity can be seen as an outcome of *aquabiopolitics*, as human governance of marine environments affects various underwater life forms, often in negative ways (Rodineliussen, 2024).

Inadvertently, aquaculture *endangers* sea cucumbers, trapping them in confined spaces where they cannot avert danger. Regardless of whether the sea cucumbers in Pemba died from poison, disease or desalination, they could not escape. Fenced into the pen, they were trapped, unable to escape the danger that ended their lives.

On 17 May 2024, we visited a farm in Unguja, Zanzibar that had lost 2,000 sea cucumbers due to extreme weather. The farm was run by a women's association. We were directed to the group by an elderly man who had participated in our workshop on 15–16 May. Impressed by our candor, he insisted we visited the women's group, which had encountered substantial problems. When we arrived, there were dozens of women waiting for us in an open hall on the seashore, a landing site for fishermen. Apparently, the group had lost all their sea cucumbers in a recent cyclone, named *Hidaya*, which struck the East African region in early May. It was an unusual weather event in this area, which rarely experiences cyclones. In Tanzania, the islands of Unguja (Zanzibar) and Mafia had been particularly hard hit. When the women came to check their farm, they found dead sea cucumbers washed ashore, even a few kilometers away. They collected them and took pictures. The Chairperson of the group explained the damage to us, while a member showed pictures on her phone (Figure 3.5):

> To give you an idea of the loss caused by *Hidaya*, it is big. We have constructed a farm for TZS seventeen (17) millions. We have added 2,000 Jongoo and we expected to sell each jongoo for [TZS] 7,000. Two thousand jongoo times seven thousand (2,000 × 7,000) it is 14,000,000. We were warned not to go to the ocean on those days, but when the cyclone hit, we came to see our Jongoo. On the 6th [of May] when we came, we saw them dead. With all the challenges, the total of the loss is TZS 34,600,000.
> (Interview transcript, Unguja, 17 May 2024).

Although their human farmers are not to be blamed for natural disasters, aquaculture can clearly *imperil* the very lives of sea cucumbers. Unable to move themselves to safer environments, farmed sea cucumbers become like sitting ducks when disasters such as cyclones or heavy rains hit their ocean

Figure 3.5 Mobile phone showing image of dead sea cucumbers. Photograph by author.

pens. If lessons from the Indian Ocean tsunami of 2004 are anything to go by, animals seem to have a knack for sensing and averting such dangers. Or they may simply raft along, moving to other places and continents, as noted during the 2011 tsunami in Japan. But these reactions rely on freedom of movement.

Nowadays, environmental changes due to climate change are amplifying the risks of aquaculture, since extreme weather can wreak havoc with sea cucumbers in captivity. Climate change impacts on aquaculture range from rising ocean temperatures, sea level rise, ocean acidification and diseases, to changes in rainfall patterns and sea surface salinity (Maulu et al., 2021). While impacts vary depending on "geographical areas, economy, climatic zones, production systems, and cultured species", climate change is expected to have "higher effects on producers in developing nations and poorer economies" as well as "small-scale farmers" (Maulu et al., 2021, p. 4). Scholars have started investigating the impact of climate change and other environmental stressors on sea cucumbers, especially salinity and temperature (Mohsen et al., 2024). Highlighting the need for more research, while noting that "sea cucumbers are significant in global fisheries and aquaculture, contributing to livelihoods and economies, particularly in coastal communities", the scholars caution that the "stress response of sea cucumbers might influence the sustainability of fisheries" (Mohsen et al., 2024, p. 17).

Whispers of the Sandfish

In waters warm where soft tides run,
The sandfish greets the rising sun.
A quiet dancer in the deep,
It tills the sands where seagrass sleeps.

But seasons shift, the heat now stays,
The ocean sings of troubled days.
Its breath grows thin, its voice runs dry,
As storm-winds carve the weeping sky.

The farmer walks the ebbing shore,
His nets once full, now hold no more.
The pens are torn, the young are few,
The sea, once kind, turns strange, untrue.

The larva drifts, too weak to fight,
Against the dark, against the night.
The waters sour, the balance fades,
As acid stains the world it made.

Yet hands rebuild, yet hearts remain,
Through salt and storm, through loss and rain.
New pens are tied, the grass restored,
A future fought for, not ignored.

For though the tides may rise and fall,
The sandfish waits, it knows them all.
And those who keep its whispers near,
Will guide its path through change and fear.
                    (https://chatgpt.com/)[2]

Notes

1 ChatGPT prompt: How does *Holothuria scabra* experience being farmed in ocean pens? 6 December 2024.
2 ChatGPT prompt: Poem on how *Holothuria scabra* is impacted by climate change in artisanal aquaculture, 12 March 2025.

4 Blue Expansion

Some Humans Think We Are Marine Gold

Figure 4.1 Sea cucumbers being dried for export. Photograph by author.

DOI: 10.4324/9781003645054-5

This chapter has been made available under a CC-BY-NC-ND 4.0 license.

The Tale of the Golden Sea Cucumber

Once upon a time, in the warm, blue waters of the Swahili Coast, where dhows with white sails danced upon the waves, there lived a humble fisherman named Juma. He was no ordinary fisherman, for he knew the ocean like the lines of his own hands. He spoke to the waves, listened to the whispers of the wind, and understood the secrets hidden beneath the sea.

One day, as Juma waded through the shallows, his bare feet sinking into the golden sand, he noticed a peculiar creature resting on the ocean floor. It was soft yet firm, dark and speckled, and when touched, it curled slightly as if it held ancient wisdom. This, he soon learned, was the sea cucumber – a creature of mystery and magic, said to bring great fortune to those who possessed it.

Juma took his discovery to the bustling markets of Zanzibar, where traders from distant lands gathered. Among them were merchants from the East, with long robes and curious eyes, who spoke of a kingdom far away – China – where sea cucumbers were more valuable than pearls. "They are treasures of the Emperor's table", one merchant told Juma, "believed to grant strength and long life. We call them bêche-de-mer, *and the Middle Kingdom craves them".*

And so began the great trade.

Fishermen like Juma scoured the ocean floor, gathering sea cucumbers in baskets, drying them under the golden African sun, and sending them off on grand dhows bound for the East. Across the vast Indian Ocean, through the monsoons and the quiet of starry nights, the sea cucumbers traveled, carried by winds and the dreams of men.

When they reached China, they were welcomed with reverence. Healers whispered of their powers, noble families served them in soups, and emperors dined upon them, believing they held the secret to vitality. The demand grew, and so did the trade.

Years turned into centuries, and the humble sea cucumber became a bridge between two worlds – the Swahili Coast and the great empire of China. But as the years passed, the ocean whispered warnings. The cucumbers, once abundant, became fewer. Fishermen ventured deeper, seeking them in hidden reefs, but the sea, wise and ancient, knew balance must be restored.

Today, the tale continues, but with a lesson – one that Juma and his people once knew well. The ocean's gifts are precious, but only when taken with respect. And so, as the waves continue their timeless dance, the story of the golden sea cucumber lives on, reminding us of the bond between distant lands and the wisdom of the sea.

The End.

(https://chatgpt.com/)[1]

It is perhaps no coincidence that ChatGPT casts it as *the golden sea cucumber*, when asked to tell the history of sea cucumber trade from Tanzania to China. While prompted to tell the story as a fairy tale, there is more to be said for the gold metaphor. Not only has the sea cucumber been valued in Chinese culture for thousands of years, priced higher than gold, but more recently it has also been talked about as *marine gold*. The notion of marine gold captures the profitable trade in bêche-de-mer during the expansion of neoliberal capitalism. But intensified commodification has resulted in a precarious situation, for sea cucumbers and coastal communities, highlighting some socioecological impacts of the Capitolocene.

This chapter will trace the commercialization of sea cucumbers through a deep history of transoceanic trade between Tanzania and China. The ocean environment where sea cucumbers live has played a critical role in the history of the Swahili civilization, connecting land and sea, while mediating transoceanic interaction. Scholars have stressed that Swahili culture is all about living with the coastal environment, which has accommodated various human activities, from agriculture to fishing and trade (Kimambo et al., 2017). The coast has also served interactions with inland communities as well as societies and cultures across the ocean. Moved by monsoon winds, these transoceanic connections have spanned African, Arab and Asian countries, shaping a cosmopolitan Swahili civilization (Sheriff, 2010). We will now anchor our sea cucumber stories in the deep history of this ocean environment, tracing early trade of exotic animals during Roman and Chinese Empires, before dwelling on bêche-de-mer trade during the expansion of Euro-American capitalism.

Nowadays, sea cucumber trade is entangled in the global blue economy and its rhetoric of blue growth, with aquaculture talked about in terms of blue revolution. In the relentless pursuit of blue growth, sea cucumber aquaculture has emerged as a technofix to problems of overexploitation of marine resources, while maintaining transoceanic trade in bêche-de-mer. In addition to wreaking havoc with marine ecosystems, the expansion of capitalist extraction into the marine frontier has resulted in problems of ocean grabbing and elite capture (Lien, 2024; Táíwò, 2022). But the dictates of the state-supported blue economy are not always adhered to, and dispossessed members of coastal communities may rely on different ways of resisting its enactment, signaling the importance of more sustainable alternatives, as argued in scholarship on blue growth (Brent et al., 2020; Ertör & Hadjimichael, 2020). As for the sea cucumbers, even AI can figure out the challenges they face in the Capitalocene.

A Deep History of Transoceanic Trade

In the open-air museum *Kaole Ruins*, which predates the Kaole sea cucumber farm by several decades, visitors can meander among coral stone buildings dating back to the 13th–15th centuries, material remnants of early

Swahili settlements. The ruins include two mosques, tombs and graves, in which local chiefs and Arab visitors were buried. Near the ruins is an area marked by a sign as the old port, now covered by mangroves. In a small museum building, some artifacts are on display, pieces of porcelain and pottery, evidence of early transoceanic trade with Arab countries and China. Some paintings on the wall depict how Kaole people have lived with the ocean, women collecting seashells and men fishing with different techniques. A model wooden fishing boat is also on display, with its sail raised. Two stands with glass tops display mud creepers (*tondo*) and cockles (*kome*), with short textual explanations of their value in local culture, as sources of animal protein and as traded objects. There is no mention of sea cucumbers, yet they have also been part of local history.

Present-day export of sea cucumbers to China can be appreciated in the historical context of this longue durée of transoceanic trade, which has included animal products that are highly valued, especially ivory and rhino horn, later bêche-de-mer. Based on archaeological findings, historians argue that trade between Swahili communities and other societies dates back to the Neolithic period, since at least 3,000 BC (Chami, 2009, pp. 203–205). Scholars have also consulted Greco-Roman documents, which show commercial contact between the East African Coast, the Middle East and the Mediterranean (Kimambo et al., 2017, p. 38).

The export of ivory and rhino horn was initially targeted at the Roman Empire. By the 2nd century AD, Tanzania was part of transoceanic trade in the Roman economic system (Kimambo et al., 2017). The East African coast was called Azania, with the metropolitan city of Rhapta as its southernmost port. In 2013, the underwater remains of the lost city of Rhapta were discovered off the Mafia Island, near the Rufiji delta in Tanzania. The most important exports from Tanzania were exotic animal parts, "especially ivory, rhinoceros horn, and tortoise shell" (Kimambo et al., 2017, p. 38). When the Roman Empire declined in the 3rd century, the market for these luxury products dropped.

Islamic countries offered an expanding market from the 7th century, alongside China and other Asian countries, and the trade in luxury products continued. Arabs, Persians, Indians and Indonesians bought exotic goods such as ivory and tortoise shell, as well as spices, incense and gold (Kimambo et al., 2017, pp. 43–44). Slave trade developed as well, with foreign traders procuring slaves by capturing or abducting local people. Imported items to Tanzania included Islamic pottery, Indian beads and Chinese porcelain.

By the 14th century, the Swahili coast had become integrated into the Islamic world, while relationships with China continued, as documented by travelers such as Ibn Battuta, Marco Polo and Chinese visitors. In the early 15th century, the Empire of China undertook sophisticated expeditions to explore what the Chinese called the Western Seas of their Middle Kingdom (Sheriff, 2010, p. 294). China could have become a colonial power, but did not. It has been suggested that despite its Sino-centric view, which saw everyone else as foreign barbarians, China was more interested in asserting its centrality and exporting its culture (Sheriff, 2010, pp. 298–299). In East Africa, China exported porcelain, silk and perfumes, while importing exotic animals, including giraffes, along with ivory and rhino horn.

In Tanzanian history writing, the arrival of Vasco da Gama in 1498, followed by various European powers, changed transoceanic interactions into more violent and oppressive relations, culminating in colonialism and imperialism. Historians underline the brutality of the Portuguese campaign for *Christianity and Commerce*, as the pre-existing system of trade was destroyed and towns along the coast were plundered (Sheriff, 2010, pp. 310–311). Indeed, as the Portuguese garrisoned major Swahili cities, trade and economic prosperity of coastal towns declined drastically "during the period of Portuguese occupation in the 16th and 17th centuries" (Kimambo et al., 2017, p. 83). While the Portuguese Empire paved the way for European dominance of maritime trade, the Chinese Empire finished its oceanic expeditions and dismantled its navy, shifting attention to domestic affairs instead. Meanwhile, Arab states continued their engagements in East Africa, not least in Zanzibar which came under Omani rule in 1698, and remained so for almost two centuries. With time, the aggressive expansion of European powers brought the Swahili coast and other littoral societies into the folds of Western capitalism and imperialism.

Before and during the colonial period, trade in sea cucumbers became entangled in the global spread of Western capitalism, alongside continued trade with China. As discussed in chapter one, Euro-American powers entered sea cucumber trade in Southeast Asia (Indonesia and Philippines) as well the Pacific (Hawaii, Tahiti and Fiji), to access the Chinese market (Wolf, 1982, pp. 258–259). From the early 1700s, Europeans competed in maritime commerce in the Indian Ocean, and by the 18th century, it was dominated by England (Campbell, 2019, pp. 177–178). This transoceanic trade included bêche-de-mer. As an "inter-trading commodity", bêche-de-mer became one of the cargoes destined for China, collected at trading centers around the Indian and Pacific Oceans, after being bought in various coastal places, including Dar es Salaam and Zanzibar in Tanzania (Sachithananthan, 1994, p. 106). In East Africa, the trade in marine products, which also included shark fins and dried cuttlefish, was dominated by Chinese traders: "Chinese merchants based in Aden bought these goods in large quantities and shipped them to China and to other Far Eastern ports." (Sachithananthan, 1994, p. 106). The withdrawal of colonial rule after the Second World War brought a "radical change in the trading pattern" (Sachithananthan, 1994, p. 107).

This deep history of transoceanic trade can be connected to scholarly analyses of the capitalist world-system (Wolf, 1982) and world-ecology in the Capitalocene (Moore, 2016a). Wolf noted how in opening up new routes for European ships in the 1400s, the Portuguese "acted as an advance guard of the European thrust", and in their search for resources abroad, "Europeans sought to control the oceans and to oust their competitors" (1982, p. 129). He insisted that capitalism developed later, with the industrial revolution in England in the 18th century as the "breakthrough from mercantile domination to the capitalist mode of production" (Wolf, 1982, p. 266). But Moore insists on dating capitalism to 1450, arguing that "early capitalism's *technics*—its crystallization of tools and power, knowledge and production—were *specifically organized* to treat the appropriation of global nature in pursuit of the endless accumulation of capital" (2016b, p. 110, emphasis in original). By mobilizing the "forces of nature" as "forces of production", capitalism as *world-ecology* has treated "the work of nature as a 'free gift'", ever since "the global conquests that commenced in 1492" (Moore, 2016b, pp. 111–112). As discussed in chapter one, European interventions into the sea cucumber trade are a telling example of these developments and the environmental effects of early commodification.

Postcolonial Exploits of Bêche-de-Mer

After independence from colonial rule, the bêche-de-mer trade between China and Tanzania continued, reaching a peak toward the end of the 20th century, when global stocks started dwindling. The post-World War period coincided with the post-revolution period in China, a time of significant political transformations. Initially, the People's Republic of China banned the import of bêche-de-mer and other commodities, which resulted in a lull in the trade until the early 1970s (Sachithananthan, 1994, p. 107). Meanwhile, Singapore and Hong Kong developed into major trade centers, with India, Indonesia and Sri Lanka as main suppliers, alongside sporadic supplies from the Pacific, Middle East and East Africa (Sachithananthan, 1994, p. 107).

From the 1980s, demand from China grew rapidly, due to its rising middle class. Sea cucumbers, as well as shark fins, held considerable appeal as luxury items, especially among Chinese consumers, not least due to "culturally rooted perceptions of class and health" (Eriksson & Clarke, 2015, p. 164). The four largest sea cucumber producing countries from 1996 to 2011 were Indonesia, Philippines, Papua New Guinea and Fiji, with trade networks often shaped by the expatriate Chinese diaspora (Eriksson & Clarke, 2015, pp. 166–167).

The boom in global sea cucumber trade from the 1980s was also evident in Tanzania. Sea cucumber fishery was carried out in the Zanzibar archipelago as well as in the mainland, especially Tanga, Bagamoyo, Dar es Salaam and Mtwara, along with Mafia Island and the Songo Songo archipelago (Mmbaga & Mgaya, 2004; Mmbaga & Mgaya, 2007; Semesi et al., 1998). Collection methods were small-scale and low-tech, through hand-picking or free diving, sometimes through scuba diving. Although no official records were kept of fishers or their catch, at the time scholars noted that in Kaole village "about

56 people are involved and about 250 kg of dried sea cucumbers are harvested per month" (Semesi et al., 1998, p. 641).

In the 1980s and 1990s, the export of sea cucumbers was substantial. Records show that exports increased from under 200 mt (metric tons) annually in the 1980s to 617 mt in 1992 (Mmbaga & Mgaya, 2007, p. 51). Import statistics from Hong Kong in the first quarter of 1996 showed that Tanzania topped the list of countries from Western Indian Ocean: Tanzania (73.8 mt), Madagascar (40.3 mt), Mozambique (23 mt), Kenya (7.4 mt). But the bêche-de-mer from Tanzania fetched a much lower price than from Madagascar, which was valued about five times higher, due to differences in processing and overall quality. Over time, the whole region experienced reduced stocks. Production in Tanzania dropped from over 1,500 mt in 1994 to 10 mt in 2004, and the number of exporters decreased as well (Mmbaga & Mgaya, 2007, p. 54). Even so, in the period 1996–2011, Tanzania was the 12th producer of sea cucumbers, out of 40 countries around the world, and only Madagascar (listed as 6th) was a greater exporter in Africa (Eriksson & Clarke, 2015, p. 167).

In 2006, the mainland government issued a ban on sea cucumber fishery, due to reports on depleted stocks as well as international guidelines. A few years later, in May 2010, IUCN listed *Holothuria scabra* as an endangered species, following a 50% decline since the 1960s–1980s (Hamel et al., 2013). In Tanzania, depleted stocks had been reported in several studies (e.g. Eriksson et al., 2010), which is why the government implemented a ban on sea cucumber fishing on the mainland, but not on Zanzibar (Mmbaga & Mgaya, 2007, p. 53). The depletion of sea cucumbers was largely attributed to collection by fishers. But fishers were not the only ones to blame. Industrial trawling of prawns by foreign vessels had sea cucumbers as bycatch (Semesi et al., 1998, p. 639). This bycatch was by no means accidental: "trawling by commercial harvesters of prawns in the 1970s was clandestinely fishing for sea cucumbers" (Mmbaga & Mgaya, 2004, p. 194). It is not known how much this clandestine bycatch amounted to, but it could have been substantial. By comparison, the daily catch of fishers varied between 5 and 55 specimens, according to a study in Songo Songo and Mafia (Darwall, 1996, cited in Mmbaga & Mgaya, 2007, p. 51).

Life Stories of Sea Cucumber Fishers and Traders

While scholarly studies offer insights into sea cucumber fishery in Tanzania, it is worth listening to the voices of people who have been involved in this practice. Their life stories offer vivid accounts of what sea cucumber fishery has meant to local people. Even after the official ban on sea cucumber fishery, the practice has continued, albeit on a much smaller scale. It is still considered somewhat profitable, providing quick cash in times of need. Sea cucumber fishery also has its own myths of a past billionaire based in Mtwara. We will get to know his story as well, but let us start with some of the people I have met in person.

An older fisher in Bagamoyo, we can call him Bwana, started fishing for sea cucumbers in Mtwara in 1981. Bwana was born in Lindi in 1967 and recounted his experiences:

> I started in Mtwara in 1981, then in same year moved to Mozambique for two months and returned to Tanzania. In 1982, I went back to Mozambique and then returned home to Lindi. In 1984 I moved to Kilwa. Also in Kilwa I continued to dive for jongoo and then I moved to Dar es Salaam in 1988 for fishing. Then a trader took me to Bagamoyo in 1989 to dive for jongoo.
>
> He went on explaining how the trader's (tajiri) agreement with divers was arranged.
>
> He would tell us there is a camp for jongoo fishing somewhere and I need about five or six fishers/divers to work for me. He then took twelve (12) of us to Gogo Hotel beach camp in Bagamoyo to dive for jongoo. He hired for us two (2) vessels (Ngalawa) that we used to go to Mwambakuni (Sandbank) to dive for jongoo and sometimes octopus.
>
> (Masimbi, interview notes from 4 July 2024)

The trader paid the divers according to their catch. His prices varied, between 5,000–10,000 for grade A and around 5,000 for grade B. Bwana used to make 40,000–60,000 in a day. They would stay in the camp for a month or so and then the trader would return to Dar, while the fishers would stay on. They would deal with other traders coming to Bagamoyo from Dar es Salaam or Zanzibar. Sometimes Bwana would process the catch and take it to the Chinese buyers himself. He would cook it and when he had about 60–80 kg, he would take it to the business areas Kariakoo, Morocco and Msasani in Dar es Salaam. He did that for a year or so, but then he just sold to local traders instead.

When we asked Bwana if he had noticed any changes in the number of sea cucumbers, he reflected: "There is a difference, there is many *jongoo* in the ocean but we don't get as much as we had". When we asked him why, he said: "Eeh, you know the eras are different. Back in the days not many people endured the hardship of working in the ocean, now we are many and the creatures must decrease".

We also talked with a local trader, we can call him Juma, who had been in the jongoo business since 1980. He had started off as a collector, but switched to trading since it earned him more money. Collection is done year-round, but especially in October to November/December and April to June, he explained to us. Juma would buy from fishers and divers, paying about 8,000 for

Grade A and 5,000 for Grade B and Grade C to E at a negotiable 3,000–5,000. He then processed the sea cucumbers by removing the stomach and washing them. He stored them in salt for a week, up to one month, until he had enough to take somewhere to sell. Before selling them, he boiled and dried them. He transported the processed sea cucumbers in plastic bags. He used to sell in Dar, but when the ban was implemented, he took them to Zanzibar instead. He used to sell to Chinese traders, and the money was quite good. He would make TZS 800,000 to 1.5 million per trip as profit. With time, the business went down, and he sold to Tanzanian middlemen instead, for less profit, only some TZS 100,000–150,000 each time.

Juma's story of gradually dwindling income is telling of how the trade has developed over the years, from the boom in the 1980s to the small-scale sales since the 2010s. His initial profits were quite substantial, providing him with more cash than most fishers earn. It enabled him to send his children to school, up to university level, which is quite unusual in coastal fishing communities. Nowadays he just earns enough for the family's upkeep.

The famous *billionaire in Mtwara* offers an almost mythical success story of a golden past. It was only when we visited Mtwara that we heard of this legendary man, his story being told by local sea cucumber farmers and government officials alike. This is how his story was told by aquaculture tutors at the FETA office in Mtwara:

> The origin of sea cucumber farming in Mtwara was attributed to Abdillahi Yusuf, who became a billionaire from sea cucumbers from Mikindani back in the 1970s and 1980s. He died in 1989. He also owned the Barracuda hotel. He inspired others. He traded in sea cucumbers from this area, from Kilwa to Mozambique. Shipped 3-4 containers of dried sea cucumbers. They were picked in intertidal zones, also by divers. He had warehouses and people would come to him to sell sea cucumbers. Two of his grandchildren want to revive his business and have come for training [at FETA].
> (Fieldnotes, Mtwara, 13 November 2023)

The story of the billionaire is also mentioned in an article entitled "Sea cucumbers: Southern Tanzania's marine gold", written by Victor Kaiza, one of the FETA officers in Mtwara and published online on 26 January 2024:

> Tanzania's southern coast comprises of two regions – Lindi and Mtwara – with 405 km of coastline. These regions were once marginalised due to the difficulties of crossing Rufiji River prior to the construction of a bridge in 2007. However, the area was famous for the production of seafood, including dried sea cucumbers, which are known as *beche de mer*. The most famous person in the sector was a trader named Abdillahi Yusuf who

became one of the richest people in Tanzania in the late 1970s and 1980s through sea cucumber exports – a trade that was the main source of income for this community. However, following the trade ban in 2003 the community has been struggling economically, relying mainly on seasonal cashew nut farming and trading.[2]

An elderly sea cucumber fisher, Abuu, talked about the famous billionaire in some detail during our interview. Abuu was sharp with many interesting insights and a concern for environmental conservation, including mangrove restoration. He was now involved in a sea cucumber farm organized by a newly registered association. When we asked him what inspired them to start sea cucumber farming, he spoke of Abdillahi Yusuf:

> Abuu's father and cousin had worked for Abdillahi, collecting jongoo in the area, also by free diving, and selling to Abdillahi. At the time there were lots of jongoo and Abdillahi had agents everywhere, who collected on his behalf. Abuu had seen him when he was a child, a billionaire who lived in Mikindani, like a simple person. He inspired a lot of people. He traded in sea products: sea shells and sea cucumbers. [He points at a large decorative shell on the table in front of him in the room we are sitting in]. The sea was untouched at the time. After he died, his company died, since his children were not engaged in it. And since then, the government has banned the business, but not in Zanzibar, so traders from Zanzibar come here [to buy sea cucumbers collected in the wild].
> (Fieldnotes, Mtwara, 16 November 2023)

In Zanzibar, I was fortunate enough to meet with a Chinese trader, we can call him Charles. I was told that the Chinese traders were not keen on talking to anyone, so when the opportunity came, I insisted. We were visiting processing facilities in a small compound in a residential area on the outskirts of Zanzibar city. Our interlocutors showed us the equipment used for processing sea cucumbers while explaining the process. They also showed us dried sea cucumbers stored in sacks, ready for export by plane to China. When we arrived, a Chinese man had greeted us, but he stayed in the background. Our interlocutors explained that he was involved in the export of sea cucumbers. I mentioned I would be happy to talk with him, but they were hesitant. I insisted. They asked him, and to my relief, he agreed. I had given him our one-page information sheet in English, and to my surprise, he had translated it into Chinese through a mobile app. He had also looked me up online, and apparently, I checked out okay. Charles did not speak English, and his Swahili was a bit peculiar, so his colleague translated his words into common Swahili, which my assistant then translated into English.

Charles was in his late 50s and had lived in Zanzibar for some time. He had come to Tanzania to export seaweed and sea cucumbers to China. He recalled that the business was good initially, up to 2015 or so. Since 2020, all business went down, because of Covid. He compared the sea cucumber business in Tanzania to that in Mozambique, where his colleagues worked. In Mozambique, there were plenty of sea cucumbers, and it was easy to collaborate with the government. In Tanzania, there were not many sea cucumbers and only a small quantity was produced. It was not easy to produce more because it was not good business, he reflected. If there would be an MoU between the Chinese and Tanzanian governments, business would be good, he suggested, but it has not happened yet. I asked him about demand in China. He answered that in China, the market was big, very good. He compared the size of populations. The population of all of Africa was about half of the population of China. He exported to China directly. In Hong Kong, the demand was only for large ones, but in China for all sizes. The sea cucumbers were exported by ship or plane; by ship if they had many tons, lower quantities were sent in small packets by plane.

These stories confirm that sea cucumber fishery has been a profitable business in Tanzania, but it has dwindled over time. According to the government and scientists, this is mainly due to the depletion of stocks, which fishers and traders have also noticed. But the reasons for this depletion are not altogether clear. The government and scientists blame fishers for overexploitation, but the problems of clandestine fishing by foreign vessels trawling for prawns are not given much attention. Meanwhile, no one challenges the market forces, as if the trade in bêche-de-mer is beyond questioning.

Neoliberal Expansion and Sea Cucumber Aquaculture

Contemporary trade in sea cucumbers can be understood in relation to neoliberalism, which can be approached as a world system and/or a mode of governing (Morningstar, 2023). While neoliberalism denotes the current phase of global capitalism, this market-driven system takes somewhat different shapes in different places depending on the historical and cultural specificities it becomes entangled with. This is why it is worth recognizing cultural hybridity in both policies and practices of neoliberalism(s). After all, the political leaders identified as the most influential proponents of neoliberal policy were located in very different parts of the world: Margaret Thatcher, Ronald Reagan, Augusto Pinochet and Deng Xiaoping (Harvey, 2007, cited in Morningstar, 2023, p. 2). While Thatcher and Reagan can be linked to neoliberal policymaking in global development, not least structural adjustment programs in Africa, we should also recognize how Deng Xiaoping navigated reforms in the rising global power of China. In aid-dependent Tanzania, the impact of structural adjustment programs has been profound. In China, integration into a neoliberal world order has not entailed a full adaptation of neoliberal policy paradigms. In both cases, we are dealing with complex hybrid constellations, in a dynamic world order of shifting powers.

In Tanzania, the surge in sea cucumber exports in the 1980s and 1990s can be linked to friendly relations with China, which were enhanced in the post-independence period. In the 1960s and 1970s, the countries were ideologically compatible, endorsing socialism rather than capitalism, in the case of Tanzania through the adoption of the Arusha Declaration in 1967. Socialism entailed internationalism, and both countries were involved in various collaborations between Asia and Africa, as well as the non-aligned movement (Nyerere, 2011). Even before independence in 1961, Tanganyika participated in the Bandung Conference in 1955, in which China was a major actor. Under the leadership of its first president Julius Nyerere, Tanzania extended its friendship with China, receiving support through financial aid and infrastructure development. One of the main outcomes of these relations was China's construction of the Tanzania-Zambia Railway (TAZARA) in the early 1970s (Qiang, 2010). But the friendship was also articulated in cultural exchange. When Chou Enlai visited Tanzania in the mid-1960s, he offered Nyerere a gift of training in acrobatics in China, which resulted in acrobatics becoming integrated into the national cultural repertoire (Uimonen, 2012, p. 145). Meanwhile, in China, African studies were developed at several universities, to deepen cultural knowledge (Anshan, 2010). Relations of friendship between the two countries were thus economic, political and cultural.

The lucrative export of sea cucumbers to China in the 1980s and 1990s coincided with Tanzania's economic crisis, followed by the adoption of neoliberal policies, enforced by the IMF and World Bank. After its experimentation with ujamaa (African socialism) in the 1970s, Tanzania experienced a severe economic crisis from 1980 to 1985 (Kimambo et al., 2017, p. 190). Realizing that his socialist policies had failed, Nyerere stepped down in 1985, following which the government agreed to "embrace neo-liberalism lock, stock, and barrel" (Shivji, 2006, p. 9). Tanzania underwent a stringent structural adjustment program dictated by the World Bank and IMF, resulting in market liberalization and multiparty politics as well as drastically reduced social development and a slimmed-down state apparatus. Tanzanian historians have reflected on the country's continued economic dependence after independence in terms of *neocolonialism,* arguing that *multilateralism* made this "exploitative apparatus more oppressive", while reducing political independence to *flag independence* (Kimambo et al., 2017, pp. 193–194). Neoliberal governance included market-oriented public sector reforms of the state apparatus. For example, the government proudly referred to Thatcher's reforms in the UK in the early 1980s as inspiration for its Public Service Reform Programme, which transformed select government agencies into more business-like executive agencies (Uimonen, 2012, pp. 69–70). Even so, the adoption of neoliberal policy did not translate into a neoliberal state, since governance in postcolonial Tanzania takes place in a hybrid *state of creolization* (Uimonen, 2012).

By comparison, in China the demand for sea cucumbers increased with its growing middle class since the 1980s, concomitant with integration into the neoliberal world system. Sea cucumber has been part of Chinese traditional medicine, but also a status symbol, enjoyed by the highest echelons in imperial China. Now it became more accessible, enabling the emerging middle classes to use their newly acquired wealth for improved health. In 1978, China embarked on a reform program, gradually opening up to and becoming a major power in the world market. Scholars have described this as an "ideological shift [...] from Maoism to economic developmentalism", which entailed a "technocratic policy agenda" and "neoliberal governmentality", yet China did not pursue neoliberal policy, but continued "disobeying the rules of the game of neoliberal globalization" (Weber, 2020, p. 1). China's post-Maoist reform program was initiated under the leadership of Deng Xiaoping, but was he really a promoter of neoliberal policy as suggested in the texts cited above? Scholars have underlined that Thatcher's efforts to educate Deng in capitalism were "doomed to failure from the start" (Mark, 2017, p. 180). During their meeting in Beijing in 1982, to discuss the future of Hong Kong, it became clear that Britain was in decline as a world power, while China was a rising power, and rather than being lectured about capitalism, the Chinese Communists were keen to learn about it in practice. Meanwhile, China insisted on state control and other non-neoliberal policies in what has been described as *market socialism*. In other words, it adopted "a hybrid governance that has combined earlier Maoist socialist, nationalist and developmentalist practices and discourses of the Communist Party with the more recent market logic of 'market socialism'" (Nonini, 2008, p. 145). In this sense, China as a whole has never been neoliberal, nor has it really pursued market socialism (Nonini, 2008).

By the early 2000s, when international organizations took an interest in sea cucumbers, Tanzania was entrenched in the neoliberal world system, albeit in a peripheral position. As you may recall, Tanzanian sea cucumber trade was initially handled by Chinese businessmen, but over time, local middlemen became involved, signaling the growth of the private sector. Meanwhile, rising demand from China's growing middle class wreaked havoc with sea cucumber stocks globally. By the late 1990s, several countries introduced restrictions on sea cucumber fishery due to depleted stocks (Mmbaga & Mgaya, 2004). With time, FAO became interested in sea cucumber aquaculture, which was piloted and researched in different countries since the 1980s, albeit with mixed results (Hamel et. al., 2022, pp. 162–163).

As exemplified by a FAO workshop on sea cucumber fishery in 2012, World Bank-driven neoliberal reform in Tanzania ushered in a donor-sponsored *workshop industry,* which included participation by local intellectual elites who embraced liberal ideology (Shivji, 2006, p. 9). In the FAO workshop in 2012, a transnational constellation of donors, intellectuals and state actors charted a roadmap that propelled the direction of sea cucumber fishery toward aquaculture, in Tanzania and other countries in the Indian Ocean (FAO, 2013). The overriding goal was the supposedly sustainable extraction of marine resources. Aquaculture seemed to offer a technoscience fix to the problem of overexploitation of sea cucumbers, while stimulating a lucrative trade with China, thus testifying to the successful spread of neoliberalism and reasserting the power of development partners in Tanzania.

Aquaculture has been the fastest growing food production technology globally since the early 1970s, surpassing wild fisheries as a source of seafood in 2015, a development that scholars have called a *global blue revolution* (Garlock et al., 2020). In salmon aquaculture, this blue revolution was spearheaded by Norway, ideologically inspired by visions of limitless growth in the ocean frontier (Lien, 2024). Globally, aquaculture is dominated by Asia (89% of global production in 2016), especially China (62% of global production), while developing countries in Africa and Latin America are considered latecomers. Over time, growth in aquaculture has not been as rapid in the Global North, not least due to environmental concerns (Garlock et al., 2020). In China, research on sea cucumber aquaculture started already in the 1950s, focusing on *Apostichopus japonicus* (Japanese sea cucumber) which also grows naturally in the region (Liu et al., 2015). Over the past decades, this aquaculture has developed into "one of the most important new aquaculture industries in China's northern coast" (2015, p. 31). In 2011, *A. japonicus* made up some 94% of sea cucumber aquaculture globally, mostly produced in China (Eriksson & Clarke, 2015, p. 166). Sold for USD 1,700–2,900/kg in Hong Kong, *A. japonicus* continues to be the most highly valued in the bêche-de-mer trade (Purcell et al., 2025, p. 5). By comparison, *H. scabra* is priced at USD 370–1,200/kg. Meanwhile, although aquaculture is touted as a sustainable alternative to satisfy demand, especially in the Chinese market, it has been suggested that "some consumers are expected to continue to prefer wild and foreign-produced sea cucumber due to concerns about contamination in local production and a desire for exclusivity" (Eriksson & Clarke, 2015, p. 168).

In Tanzania, sea cucumber aquaculture has been promoted by donor agencies, state actors and marine scholars, which makes it an exemplary case of *scientific capitalism* in a neoliberal world order (Ferguson, 2006). When discussing neoliberal governance in Africa, Ferguson conceptualizes structural adjustment programs as scientific capitalism, a demoralizing regime of economic correctness whereby "controversial and widely disputed claims are blandly asserted as simple, incontestable, scientific fact" (Ferguson, 2006, p. 78). He exemplifies his argument with the cultivation of food crops for export and the comparative advantage of pursuing niche crops in unstable agricultural export markets. Some two decades later, it would seem that sea cucumber aquaculture is framed in similar terms of scientific capitalism, now in the context of the Blue Economy. The sea cucumber is viewed as a lucrative marine resource, a niche seafood product cultivated for export in an uncertain and unregulated market. Meanwhile, aquaculture enclosures are limiting access to the ocean, thus dispossessing fishers from their livelihoods, while enticing local communities through promises of marine gold.

But no matter how powerful neoliberal governance may seem, its expansion is far from a straightforward implementation of liberal norms and values, since its global expansion is always entangled in the complexities of local and translocal circumstances. In marginalized coastal communities, state authorities tend to be treated with some skepticism, while people find ways of circumventing superimposed rules and regulations, in order to safeguard their livelihoods and their way of life, which is intricately interlinked with the ocean.

Defiant Resistance to Elite Capture and Dispossession

As much as blue growth claims to improve local livelihoods while protecting the ocean environment, it tends to enact *elite capture* of ocean space, thus *dispossessing* coastal communities of seascapes that they depend on for their livelihoods. Among the people whose life stories I have described above, only government officials and wealthy businessmen could be classified as elite; fishers tend to belong to less privileged social strata. In Swahili ocean worlds, fishers and other coastal peoples depend on the ocean for everyday life, which makes the top-down Blue Economy model rather problematic. Similar to dispossession in sea cucumber aquaculture in Madagascar (Baker-Médard & Kroger, 2023), by enclosing ocean space for aquaculture, some coastal peoples are denied access to their lived ocean environment, while investors and traders can reap the benefits of marine resource extraction.

In development studies, elite capture is a well-established concept, denoting how members of a privileged elite usurp the benefits of development efforts aimed at more impoverished populations. Elite capture has already been identified as a threat to the diversification of livelihoods in coastal Tanzania (Torell et al., 2017). In more philosophical terms, we can appreciate *elite capture* as "symptomatic of social systems with unequal balances of power", a feature of *racial capitalism*, with elite referring to a relationship rather than a stable identity (Táíwò, 2022, pp. 22–23). Globally, "elite capture is perhaps clearest at the multinational level", with the World Bank and International Monetary Fund (IMF) holding "immense governing power" over "basic features of non-elite life", including the price of food, public services and availability of jobs (Táíwò, 2022, p. 27). If anything, their power has increased since the enforcement of structural adjustment programs, and the continued implementation of neoliberal public policy, technocratic governance and scientific capitalism (Ferguson, 2006). In the Blue Economy, elite capture is enacted in various forms.

In sea cucumber aquaculture, elite interventions stretch from local to global levels, from affluent members of local communities, through business investors and government officials to global traders and middle-class consumers. It is embedded in and enabled by hierarchical social relations, with farmers (producers) at the lowest rungs, directed by local leaders (harbingers), facilitated by loans or businesses (investors), through government regulation (authorities), driven by development assistance (donors), for the global market (traders), to meet the demand from a growing Chinese middle-class (consumers). To satisfy the interests of these elite actors, the ocean is enclosed and privatized, a sneaky form of ocean grabbing enacted and guarded by hopeful farmers, in their new role as blue entrepreneurs tasked with the tedious management of what is promised to be marine gold.

Even so, at the local community level, relations between farmers and elite tend to be rather complex. In most coastal communities, the investment required for sea cucumber farming is well beyond the means of an average household. In some cases, local sponsors step in to initiate the business, financing the initial infrastructure and stock, as a soft loan. For instance, in Kaole, a local businessman sponsored the initial setup. As a wealthy person, he is referred to as tajiri (rich), which is a term that is also used for prosperous traders. These relations between wealthy sponsors and local fishers should not be understood as patron-clientism, but as part of more complex fisher-trader relations, which include but also go beyond economic transactions and credit provisions (Drury O'Neill & Crona, 2017). To members of the sea cucumber association, this sponsor is spoken of in positive terms, as someone who helped them. By contrast, leaders of the association have often been criticized and accused of theft. This could be indicative of how fishers perceive and relate to state authorities in general, since the association is run by the Beach Management Unit (BMU). But it is also telling of the more well-to-do socioeconomic status of some of the leaders, who originate elsewhere and whose social position entails a certain social distance from local fishers and divers.

Elite attitudes and relations can also be appreciated in the context of technocratic governance, especially between experts/officials and fishers/locals. Experts and officials often talk about the need to educate local people about the blue economy, to raise their awareness and build their capacity, as they express it, thus articulating common buzzwords in development lingo. While this jargon of capacity development captures the *racial vernaculars* of development (Pierre, 2020), it also reveals its *elitist composition*. As much as members of the educated elite played a central role in the struggle for independence from colonial rule, many members of the "African leaders of the intelligentsia" were "capitalist, and shared fully the ideology of their bourgeois masters" (Rodney, 2012, p. 279). Within years of independence, "a new class of wabenzi was building up", and "newly acquired state positions were used to accumulate wealth, flaunted in conspicuous consumption" by some politicians and civil servants (Shivji, 2006, p. 4; cf. Bayart, 1989).[3] A few decades later, a "neo-liberal middle class" developed, while the "neo-liberal intellectual elite" dissolved into the neoliberal political elite or business elite, or the world of donor-funded NGOs and consultancies (Shivji, 2006, p. 11). As much as Fanon warned against the nationalist bourgeoise (2021 [1961]), and despite Nkrumah's premonition of the dangers of neocolonialism (1963), an internally diversified and transnationally oriented cosmopolitan elite has been integral to neoliberal entrenchment, in Africa and throughout the world at large.

Let us return to the issue of theft and security in sea cucumber farming, to probe how elite capture in aquaculture may be resisted by dispossessed fishers and divers. As noted around the world, security is of primary concern in sea cucumber aquaculture, since theft is a recurring challenge (Hamel et. al., 2022). It is very difficult to guard open water ocean enclosures, some of which can be quite large. Lien (2015) has discussed how borders in industrial salmon aquaculture in Norway reconfigure biosocial relations, separating nature/culture, wild/domestic. In artisanal aquaculture, the fenced-in borders of sea cucumber farms erect a boundary between humans, especially between farmers/investors and divers/fishers. As classificatory devices, these boundaries enact the neoliberal logic of the blue economy, controlling marine resources (sea cucumbers) and their exploitation (aquaculture) through select human actors (blue entrepreneurs).

Through the idioms of security and theft, elite capture of ocean space is both enacted and resisted. For the farmers, security entails control and authority over ocean space, assuring the legality of their aquaculture practices. For fishers, the fences are an obstruction, hindering them from entering areas in the vast ocean, no matter how often they have traversed or fished there in the past. In this struggle over ocean space, ownership is concealed in the rhetoric of security, while fishers pose a threat to ocean grabbing, thus challenging both morality and legality. As recalled by the managers of a large commercial farm in Zanzibar:

> The challenges that exist right now, first of all, is the lack of understanding among people around us, the community that surrounds us, including these villagers and those fishermen who come here to deal with their fishing business. Little understanding. We try to educate them. We stay with them, so they understand gradually. We can achieve it. Maybe we do it in front of them. We spend a lot of time going around with the boat. Someone who is not allowed here, even though he has seen the buoys, he is coming in. So, we catch him, educate him, take him out and let him stay outside. This is a little challenge we have right now.
> (Interview with investor in Zanzibar, May 2024)

What is described as lack of understanding can be understood as acts of defiance, as fishers refuse to respect borders erected in the ocean, thereby counteracting ocean grabbing. Far from being ignorant, they understand that borders signify privatization of ocean space, curtailing their access to the sea that sustains them. But do they need to respect such acts of ocean grabbing? Especially when the profits of this farm of 60,000 square meters (almost 15 acres) only go to a small group of investors, none of whom are from the local community.

What is legally correct is not always morally justifiable. A fisher who had been caught picking sea cucumbers outside a farm felt that it was *unfair* that he was accused of stealing. After all, he picked them outside the borders. When we asked him why the government had banned sea cucumber fishing, he articulated an interesting interpretation of government policy:

> It has been banned, because it is a sea natural resource for export, it is like gold. The government banned it because they wanted it to be organized and done by communities and be harvested at specific periods with license, so that it becomes profitable to many people and generates more income than selling it little by little and make small profits for only some individuals.
>
> (Summary of interview in Kaole, June 2022).

Meanwhile, illegal collection and trade in sea cucumbers continue, albeit on a smaller scale. Cognizant of the ethical obligation to do no harm, I will not dwell too much on this, but suffice it to say that some fishers and divers continue to collect sea cucumbers in the wild, sell them to local traders, often raw, straight from the sea. The traders do some processing before selling them onward to exporters. While prices for collected sea cucumbers are lower than for farmed ones, they are paid cash-in-hand. As a motorbike driver explained to us, when the season is right and he needs some cash, he simply walks into the ocean and collects some sea cucumbers. Sometimes traders sit on the seashore waiting for divers, whose catch may contain some sea cucumbers. Typically, the number each diver catches is no more than a handful or a dozen, of varying sizes. Hardly enough to endanger the species.

Circumventing aquaculture, sea cucumber fishery can be interpreted as yet another act of defiance, a refusal to follow the state-sanctioned market, while engaging in more informal, albeit illicit, global trade. The two forms of trade are interlinked, given the limited number of traders, not to mention exporters. Meanwhile, the inability to enforce the ban on fishery has multiple reasons. One problem is tracing. Physiologically, there is no way to determine origin. Whether collected in the wild or from farms, the sea cucumbers look alike. Another problem is the tendency among some officials to overlook illicit trade, as long as some of the profits are shared. The seemingly endless demand from China poses another problem: there is always a buyer for sea cucumbers. And while the dream of huge earnings keeps farmers going, fishers and divers seem rather content with some quick cash.

Paradoxically, illicit fishery enacts the neoliberal ideology more systematically than state-sanctioned aquaculture. Sea cucumber fishery is an individualistic endeavor that maximizes profit, while minimizing investment, in money, time and effort. Fishers earn quicker cash, and when measured against their input, the financial return is far more beneficial. Aquaculture by contrast reflects a hybrid form of neoliberal governance, with interventionist state actors securing ocean space for entrepreneurship, while financial benefits are left to an unregulated global market that profits from licit as well as illicit trade in endangered seafood.

The impossibility of fully restricting sea cucumber fishery, or to maintain socioecological balance through aquaculture, points to the limits of human control over blue growth. As much as aquaculture enacts "ocean-as-infrastructure" through marine monocultures that resemble "plantation economies", narrated through "oceanic hype" of limitless growth in marine frontiers, it has detrimental impacts on marine ecologies and coastal livelihoods (Lien, 2024, p. 404). As discussed earlier, aquaculture heightens the risk of disease, while making sea cucumbers more vulnerable to the effects of climate change. As discussed here, it also results in growing inequalities in coastal communities that already face marginalization.

From a degrowth perspective, this lack of control points to the inherent paradox of blue growth. One of the most fundamental contradictions of capitalism is the double bind of economic growth and socioecological sustainability (Eriksen, 2016). By reframing blue growth as blue fix, we can appreciate how aquaculture attempts to overcome the overexploitation of fisheries, in this case sea cucumber, while "ignoring the negative socioecological impacts" (Brent et al., 2020, p. 31). Such "irreparable ecological and social contradictions" show how blue growth not only "fails to address the underlying causes of environmental degradation, climate change and dispossession of coastal communities" but rather than fixing such incongruities, it actually exacerbates them (Brent et al., 2020, pp. 40–41).

Survival in the Capitalocene?

The Capitalocene *– a term that critiques the Anthropocene by emphasizing the role of global capitalism in environmental change – poses several survival challenges for sea cucumbers (Holothuroidea). These marine invertebrates, vital to ocean ecosystems, are increasingly threatened by human activities. Key challenges include:*

1 ***Overexploitation and Illegal Harvesting***

- *Sea cucumbers are heavily harvested for the Asian market, where they are consumed as a delicacy and used in traditional medicine.*
- *Overfishing, often driven by high economic demand, has led to severe population declines and the collapse of some regional fisheries.*
- *Illegal, unreported and unregulated (IUU) fishing exacerbates the crisis, with some species now endangered.*

2 **Climate Change and Ocean Warming**

- Rising ocean temperatures stress sea cucumbers, affecting their metabolism, reproduction and survival.
- Warming waters can disrupt their natural habitats, altering food availability and increasing susceptibility to disease.

3 **Ocean Acidification**

- Increased CO_2 absorption by oceans lowers pH levels, which can impair sea cucumber physiology, including digestion and reproduction.
- Acidification may also reduce the availability of calcium carbonate, which some species use for skeletal structures.

4 **Habitat Destruction**

- Coastal development, bottom trawling and destructive fishing methods damage seagrass beds and coral reefs, essential habitats for many sea cucumber species.
- Pollution from industrial and agricultural runoff introduces toxins and disrupts microbial communities critical to their diet.

5 **Microplastic and Heavy Metal Pollution**

- Sea cucumbers play a key role in sediment processing, but their filter-feeding behavior makes them highly susceptible to ingesting microplastics and toxins.
- Bioaccumulation of heavy metals from pollution can have long-term health impacts and may reduce their reproductive success.

6 **Disruptions to Ecological Roles**

- As bioturbators, sea cucumbers recycle nutrients, aerate sediments and maintain marine ecosystem health.
- Population declines threaten these ecological functions, leading to broader consequences for coral reefs, seagrass meadows and ocean biodiversity.

Conclusion

In the Capitalocene, the survival of sea cucumbers is intricately linked to capitalist-driven extractivism, climate change, and pollution. Effective conservation strategies – including sustainable fisheries management, habitat protection and stronger regulations – are crucial to ensuring their survival and the stability of marine ecosystems.

(https://chatgpt.com/)[4]

Notes

1 ChatGPT prompt: Tell the history of sea cucumber trade from Tanzania to China like a fairy tale. 7 February 2025.
2 https://thefishsite.com/articles/sea-cucumbers-southern-tanzanias-marine-gold
3 Wabenzi refers to corrupt officials, or their family members, and is derived from "wa" (people) and Mercedes-Benz cars, according to Wikipedia.
4 ChatGPT prompt: Challenges of survival for sea cucumbers in the Capitalocene. 14 March 2025.

5   Ocean Creatures Revisited
Voices from the Seafloor

Sea Cucumber Stories from a Multispecies Perspective

How time flies! The sea cucumber has been known for over 1000 years. Talking about sea cucumber is just like reading Chinese history. During this long historical period, sea cucumber has been deeply branded by the Chinese. It is all at once a potent drug in traditional Chinese medicine, a delicacy in Chinese cuisine, a favored gift and tribute, and an object of art. It is intimately linked with the life of Chinese people. What is more important is that it is just like a mysterious and beautiful natural angel that brings health and frailty to human beings, and drives people to explore knowledge and tenaciously probe into it. Keeping with tradition and history, and heartfelt ancestral ties to sea cucumbers, Chinese scholars are still studying the legendary species today. *This ancient life form, over 0.6 billion years old, is now crossing paths with human beings, but has yet to reveal all of its secrets*. All the more reason to explore and study it!

(Yang & Bai, 2015, p. 22, emphasis added)

In this scholarly appraisal of Chinese culture and history, the sea cucumber is valued in many ways, as an ingredient in traditional medicine, as a delicacy in Chinese cuisine, as a status gift and tribute and as an art object. What is missing is a valuation of the sea cucumber as an ocean creature. Even the recognition of the sea cucumber as an *ancient life form* focuses on how it is *crossing paths with human beings,* a human-centered perspective that celebrates it as a source of knowledge for humans to unravel.

Let us turn to Western marine scholars, who focus more on species taxonomies and commercial value than on culture, thus connecting science and commerce. In a recent publication on *Commercially Important Sea Cucumbers in the World*, some of the world's leading marine scholars elucidate on taxonomic details of 84 species of sea cucumber (Purcell et al., 2023). The

DOI: 10.4324/9781003645054-6
This chapter has been made available under a CC-BY-NC-ND 4.0 license.

publication is prepared for FAO, and includes information on how to process sea cucumbers, along with their market value:

> Since the 1980s, sea cucumber harvesting has boomed but many stocks have since collapsed. Towards the end of the "boom" part of a "boom-and-bust" fishing cycle, populations of some species had been reduced to such low levels that there was little capacity for natural recovery and replenishment, leading to their economic and ecological depletion. Consequently, Asian markets have increased demand, putting severe pressure on fishers in producing countries to supply markets with new species.
> (Purcell et al., 2023, p. 1).

As much as this publication acknowledges that some species have been overexploited, they still offer detailed information, for example on *H. scabra*. The species is mentioned in descriptions of processing techniques and multispecies fisheries. In a section fully devoted to *H. scabra* (Purcell et al., 2023, pp. 116–117), exploitation of the species is described in greater detail, while its endangered status is merely mentioned at the end of the section:

EXPLOITATION

Fisheries: Exploited heavily in artisanal and industrial fisheries throughout its range. Generally harvested by free diving and gleaning. It was overharvested in Papua New Guinea, Solomon Islands, Torres Strait and Vanuatu, leading to moratoria. Harvested extensively in northern Australia and throughout Southeast Asia, Africa and the Indian Ocean. In India, it was an important species until a fishing moratorium in 2001.

[…]

OTHER USEFUL INFORMATION

This species is widely reported as overexploited in many localities, and listed as Endangered on the IUCN Red List.
(Purcell et al., 2023, p. 117)

This scientific portrayal of *H. scabra* affirms its commercial value, even explaining how to process it and what the market price is, while noting that it is also listed as endangered. Through photographs and drawings, the sea cucumber is again objectified, while the scientific language provides some evidence-based framings of it as a commodity. Although the publication was funded by the FAO Fisheries and Aquaculture Division and a program on fisheries management funded by the Norwegian Agency for Development Cooperation (Norad), there is no elaboration on aquaculture. Instead, the main form of exploitation is fishery: "Fished throughout the Indian Ocean and Red Sea. Harvested by hand by gleaning and by free-diving" (Purcell et al., 2023, p. 25).

Globally, FAO is now pursuing a Blue Transformation strategy, which privileges aquaculture, while incorporating climate change as yet another challenge to be overcome. This strategy seeks to utilize aquatic food systems for food security, while buffering the impacts of climate change and contributing to poverty alleviation (FAO, 2022, p. 4). Aquaculture is the first of three pillars in this strategy, alongside fisheries and value chains, with the stated objective: "Sustainable aquaculture intensification and expansion satisfies global demand for aquatic food and distributes benefits equitably" (FAO, 2022, p. 7). Two of the targets state: "Equitable access to resources and services delivers new and secures existing aquaculture-based livelihoods" and "Aquaculture operations minimize environmental impact and use resources efficiently". The objective of value chain is "Upgraded value chains ensure the social, economic and environmental viability of aquatic food systems", with a target "Fisheries and aquaculture products access international markets more effectively" (FAO, 2022, p. 7).

When it comes to aquaculture in Africa, FAO underlines the importance of more industrial approaches instead of artisanal practices, as elaborated in its *Policies and strategic plans in support of aquaculture development in sub-Saharan Africa* (FAO, 2024):

> The reasons for the lack of success in aquaculture development in Africa have been reviewed by Brummett, Lazard and Moehl (2008), who stated that aquaculture at an artisanal level will not be profitable, but commercial aquaculture can produce significant profits. However, appropriate incentives and policies must be accompanied by non-farm models and centres of excellence to provide guidance for profitable business ventures. […]
>
> The action plan should be developed collaboratively by stakeholders from various regions, with the objective of creating an enabling environment for the growth of the aquaculture sector, especially for small- and medium-scale enterprises. The plan must focus on overcoming challenges and capitalizing on opportunities to contribute to food security, livelihoods and wealth creation across Africa.
>
> (Menezes et al. 2024, p. 3).

This FAO Roadmap reasserts the ideology of blue growth and blue economy, while promoting *industrial aquaculture* as a solution to problems of food security, livelihoods and wealth creation. Framed in the rhetoric of stakeholder collaboration as well as small- and medium-scale enterprises, it conceals the power dynamics at play, while the practices of artisanal fishers and farmers are portrayed as not profitable. No mention is made whatsoever of the socioecological problems that aquaculture raises.

All these anthropocentric appraisals fail to acknowledge sea cucumbers as ecologically significant life forms in the ongoing worlding of the ocean: a fellow creature quietly living on the seafloor, taking care of the ocean. Instead, the sea cucumber is seen through human eyes and needs, as a culturally valued delicacy, as a healing traditional medicine or as a commodity in global trade that can be mass-produced through aquaculture.

This failure to value sea cucumbers as ocean creatures is having disastrous consequences, made worse by aquaculture. Far from protecting sea cucumbers, my research over the last few years has shown that in aquaculture, sea cucumbers are decimated by the thousands, due to lack of human care as well as environmental changes beyond human control (Uimonen, 2025). For example, recall how some 10,000 fingerlings died in transportation from Kaole to Mtwara, stored in a black plastic container on a truck driven in daylight. Or how some 2,000 sea cucumbers died after the cyclone Hadiya hit Unguja island in Zanzibar, a few months after 1,400 sea cucumbers died after heavy rainfalls on Pemba Island. These sea cucumbers were all trapped in farms, unable to save themselves.

In this age of environmental crisis, we can ill afford to ignore the impact of our value systems, anthropocentric ontologies that are wreaking havoc with our multispecies worldings.

Perhaps it is time to ask whether market demand for ocean creatures like sea cucumbers should be privileged over the need to safeguard ocean environments? Or do we need to imagine another multispecies world of climate and environmental justice in the Capitalocene?

Figure 5.1 Sea cucumber in its ocean environment. Photograph by author.

A Message from the Sea Cucumbers: Protect Us, Protect the Ocean

Greetings, humans! We are the humble sea cucumbers, the quiet custodians of the ocean floor. Though we may not be the most famous or flashy creatures, we play a vital role in keeping marine ecosystems healthy.

We are nature's recyclers, cleaning the seabed by breaking down organic matter and enriching the sediment, making it more fertile for other marine life. Our presence helps maintain the delicate balance of the ocean, but we are under threat. Overfishing, habitat destruction, and pollution are pushing many of our species toward decline.

When we disappear, the ocean suffers. Without us, the seabed becomes less healthy, carbon cycling is disrupted, and biodiversity declines. The ocean is the heartbeat of the planet, providing oxygen, regulating climate, and sustaining life – including yours!

We urge you to protect marine ecosystems by reducing pollution, supporting sustainable fishing, and advocating for marine conservation. Your actions, big or small, make a difference. Protect us, and you protect the ocean – our shared home.

With gratitude from the ocean floor,

The Sea Cucumbers

(https://chatgpt.com/)[1]

Note

1 ChatGPT prompt: Message from sea cucumbers on the importance of protecting them and the ocean. 8 February 2025.

References

Ammar, N. (2019). Islamic ethics. In A. Kothari, A. Salleh, A. Escobar, F. Demaria, & A. Acosta (Eds.), *Pluriverse: A post-development dictionary* (pp. 212–213). Tulika Books.

Anshan, L. (2010). African studies in China: A historiographical survey. In A. Harneit-Sievers, S. Marks, & S. Naisu (Eds.), *Chinese and African perspectives on China in Africa* (pp. 2–24). Pambazuka Press.

Baker-Médard, M., & Kroger, E. (2023). Troubling the waters: Gendered dispossession, violence and sea cucumber aquaculture in Madagascar. *Society & Natural Resources, 37*(4), 443–470. https://doi.org/10.1080/08941920.2023.2288674

Barbesgaard, M. (2018). Blue growth: Savior or ocean grabbing? *Journal of Peasant Studies, 45*(1), 130–149. https://doi.org/10.1080/03066150.2017.1377186

Barker, K., Taylor, S., & Dobson, A. (2013). Introduction: Interrogating bio-insecurities. In A. Dobson, K. Barker, & S. Taylor (Eds.), *Biosecurity: The socio-politics of invasive species and infectious diseases* (pp. 1–25). London: Routledge.

Bayart, J. F. (1989). *The state in Africa: The politics of the belly*. Polity Press.

Bell, G. (2021). Talking to AI: An anthropological encounter with artificial intelligence. In *The SAGE Handbook of Cultural Anthropology*. SAGE Publications Ltd. https://doi.org/10.4135/9781529756449

Bhat, H. (2023). Stickiness in a monsoon air methodology. In A.O. Andersen, N. Bubandt, & R. Cypher (Eds.), *Rubber boots methods for the Anthropocene: Doing fieldwork in multispecies worlds* (pp. 200–224). University of Minnesota Press.

Blaser, M., & de la Cadena, M. (2018). Pluriverse: Proposal for a world of many worlds. In M. de la Cadena & M. Blaser (Eds.), *A world of many worlds* (pp. 1–22). Duke University Press.

Brent, Z. W., Barbesgaard, M., & Pedersen, C. (2020). The blue fix: What's driving blue growth? *Sustainability Science, 15*, 31–43. https://doi.org/10.1007/s11625-019-00777-7

Brugidou, J., & Fabien, C. (2018). AnthropOcean': Oceanic perspectives and cephalopodic imaginaries moving beyond land-centric ecologies. *Social Science Information, 57*(3), 359–385. https://doi.org/10.1177/0539018418795603

Bubandt, N. (2023). Tidalectic ethnography: Snorkeling the coral reefs of the Anthropocene. In A.O. Andersen, N. Bubandt, & R. Cypher (Eds.), *Rubber boots methods for the Anthropocene: Doing fieldwork in multispecies worlds* (pp. 171–199). University of Minnesota Press.

Bubandt, N., Andersen, A. O., & Cypher, R. (2023). Introduction. Rubber boots methods; Outline for a multispecies study of the Anthropocene. In A.O. Andersen, N. Bubandt, & R. Cypher (Eds.), *Rubber boots methods for the Anthropocene: Doing fieldwork in multispecies world* (pp. 1–35). University of Minnesota Press.

Campbell, G. (2019). *Africa and the Indian Ocean world from early times to circa 1900*. Cambridge University Press.

Chami, F. (Ed.) (2009). *Zanzibar and the Swahili coast from c.30,000 years ago*. E&D Vision Publishing Limited.

Chao, S., Bolender, K., & Kirksey, E. (2022). *The promise of multispecies justice*. Duke University Press.

Chemhuru, M. (2023). The ontological foundation of African Knowledge: A critical discourse in African communitarian Knowledge. In P. A. Ikhane, & I. E. Ukpokolo (Eds.), *African epistemology: Essays on being and knowledge* (pp. 91–104). Routledge.

Crespi-Abril, A. C., & Rubilar, T. (2023). Ethical considerations for echinoderms: New initiatives in welfare. *Animals, 13*, 3377. https://doi.org/10.3390/ani13213377

Drury O'Neill, E., & Crona, B. (2017). Assistance networks in seafood trade – A means to assess benefit distribution in small-scale fisheries. *Marine Policy, 78*, 196–205.

Dua, J. (2024). Anthropology of and from the ocean. *Annual Reviews of Anthropology, 54*, 165–181.

Eisenstein, C. (2019). Earth spirituality. In A. Kothari, A. Salleh, A. Escobar, F. Demaria, & A. Acosta (Eds.), *Pluriverse: A post-development dictionary* (pp. 157–160). Tulika Books.

Elliott, D., & Culhane, D. (Eds.) (2017). *A different kind of ethnography: Imaginative practices and creative methodologies*. University of Toronto Press.

Eriksen, T. H. (2016). *Overheating: An anthropology of accelerated change*. Pluto Press.

Eriksson, H., & Clarke, S. (2015). Chinese market responses to overexploitation of sharks and sea cucumber. *Biological Conservation, 184*, 163–173.

Eriksson, H., de la Torre-Castro, M., Eklöf, J., & Jiddawi, N. (2010). Resource degradation of the sea cucumber fishery in Zanzibar, Tanzania: A need for management reform. *Aquatic Living Resources, 23*(4), 387–398.

Eriksson, H., Robinson, G., Slater, M., & Troell, M. (2012). Sea cucumber aquaculture in the Western Indian Ocean: Challenges for sustainable livelihood and stock improvement. *AMBIO, 41*, 109–121.

Ertör, I., & Hadjimichael, M. (2020) Editorial: Blue degrowth and the politics of the sea: Rethinking the blue economy. *Sustainability Science, 15*, 1–10. https://doi.org/10.1007/s11625-019-00772-y

Escobar, A. (2020). *Pluriversal politics: The real and the possible*. Duke University Press.

Etieyibo, E. (2017). Anthropocentrism, African metaphysical worldview and animal practices: A reply to Kai Horsthemke. *Journal of Animal Ethics, 7*(2), 145–162. https://doi.org/10.5406/janimalethics.7.2.0145

Fabiani, G., Namukose, M., Katikiro, R. E., Yussuf, Y., Steinmann, N., & Msuya, F. (2023). Potential and challenges of integrated multi-trophic aquaculture (IMTA) system of seaweed and sea cucumber in Tanzania. In M. Wolff, S. Ferse, & H. Govan (Eds.), *Challenges in tropical coastal zone management: Experiences and lessons learned* (pp. 133–148). Springer.

Fanon, F. (2021 [1961]). *The wretched of the earth*. Revised sixtieth anniversary edition. Grove Press.
FAO. (2013). *Report on the FAO workshop on sea cucumber fisheries: An ecosystem approach to management in the Indian Ocean (SCEAM Indian Ocean)*. Mazizini, Zanzibar, Tanzania, 12–16 November 2012. FAO: FIRA/R1038.
FAO. (2022). *Blue transformation: roadmap 2022–2030. A vision for FAO's work on aquatic food systems*. FAO.
FAO. (2024). *Policies and strategic plans in support of aquaculture development in sub-Saharan Africa. A tool for planning and resource mobilization*. FAO.
Farnell, B. (2011). Theorizing 'the body' in visual culture. In M. Banks & J. Ruby (Eds.), *Made to be seen: Perspectives on the history of visual anthropology* (pp. 136–158). University of Chicago Press.
Ferguson, J. (2006). *Global shadows: Africa in the neoliberal world order*. Duke University Press.
Garlock, T., Asche, F., Anderson, J., Bjørndal, T., Kumar, G., Lorenzen, K., Ropicki, A., Smith, M. D., & Tveterås, R. (2020). A global blue revolution: Aquaculture growth across regions, species and countries. *Reviews in Fisheries Science & Aquaculture, 28*(1), 107–116.
Grasseni, C. (2011). Skilled visions: Toward an ecology of visual inscriptions. In M. Banks & J. Ruby (Eds.), *Made to be seen: Perspectives on the history of visual anthropology* (pp. 19–44). University of Chicago Press.
Gruber, M. (2022). *Sharing the camera. A guide to collaborative ethnographic filmmaking*. Sean Kingston Publishing.
Hamel, J. F., Eeckhaut, I., Conand, C., Sun, J., Caulier, G., & Mercier, A. (Eds.) (2022). *Global knowledge on the commercial sea cucumber Holothuria scabra*. Elsevier Science & Technology.
Hamel, J. F., Conand, C., Pawson, D. L., & Mercier, A. (2001). The sea cucumber *Holothuria scabra* (Holothuroidea: Echinodermata): Its biology and exploitation as beche-de-mer. *Adv. Mar. Biol, 41*, 129–223.
Hamel, J. F., Mercier, A., Conand, C., Purcell, S., Toral-Granda, T.-G., & Gamboa, R. (2013). *Holothuria scabra. The IUCN Red List of Threatened Species 2013*: e. T180257A1606648. https://dx.doi.org/10.2305/IUCN.UK.2013-1.RLTS.T180257A1606648.en
Haraway, D. (1998). Situated knowledges: The science question in feminism and the privilege of partial perspective. *Feminist Studies, 14*(3), 575–599.
Haraway, D. (2008). *When species meet*. University of Minnesota Press.
Haraway, D. (2016). *Staying with the trouble. Making kin in the Chthulucene*. Duke University Press.
Harrison, P. (1999). Subduing the earth: Genesis 1, early modern science and the exploitation of nature. *The Journal of Religion, 79*(1), 86–109.
Hastrup, K., Münster, U., Tsing, A., & Bubandt, N. (2022). Troubling methods in the Anthropocene: A roundtable discussion. In A. O. Andersen, N. Bubandt, & R. Cypher (Eds.), *Rubber boots methods for the Anthropocene: Doing fieldwork in multispecies world* (pp. 371–409). University of Minnesota Press.
Helmreich, S. (2011). Nature/culture/seawater. *American Anthropologist, 113*(1), 132–144.
Helmreich, S. (2023). *A book of waves*. Duke University Press.
Hofmeyr, I. (2010). Universalizing the Indian Ocean. *PMLA/Publications of the Modern Language Association of America, 125*(3), 721–729. https://doi.org/10.1632/pmla.2010.125.3.721

References

Hofmeyr, I. (2012). The complicating sea: The Indian Ocean as method. *Comparative Studies of South Asia, Africa and the Middle East*, *32*(3), 584–590.

Hornborg, A. (2020). Anthropology in the Anthropocene. *Anthropology Today*, *36*(2), 1–2.

Horsthemke, K. (2017). Animals and African ethics. *Journal of Animal Ethics*, *7*(2), 119–144. https://doi.org/10.5406/janimalethics.7.2.0119

Ikhane, P. A., & Ukpokolo, I. E. (Eds.) (2023). *African epistemology: Essays on being and knowledge*. Routledge.

Ingold, T. (2000). *The perception of the environment. Essays on livelihood, dwelling and skill*. Routledge.

Ingold, T. (2022). *Imagining for real. Essays on creation, attention and correspondence*. Routledge.

Ingold, T., & Palsson, G. (Eds.) (2013). *Biosocial becomings: Integrating social and biological anthropology*. Cambridge University Press.

Jaeger, G. F. (1833). *De Holothuriis*. Gessnerianis, Turici.

Jue, M. (2020). *Wild blue media: Thinking through seawater*. Duke University Press.

Khaldûn, I. (1967). *The Muqaddimah: An introduction to history*. Princeton University Press. Original work published 1958

Kimambo, I., Maddox, G., & Nyanto, S. (2017). *A new history of Tanzania*. Mkuki na Nyota.

Kirksey, E., & Chao, S. (2022). Who benefits from multispecies justice? In S. Chao, K. Bolender, & E. Kirksey (Eds.), *The promise of multispecies justice* (pp. 1–21). Duke University Press.

Kirksey, E., & Helmreich, S. (2010). The emergence of multispecies ethnography. *Cultural Anthropology*, *25*(4), 545–576.

Knappert, J. (1979). *Myths and legends of the Swahili*. East African Educational Publishers (E.A.E.P.).

Latour, B. (1986). Visualization and cognition: Drawing things together. In H. Kuklick (Ed.), *Knowledge and society studies in the sociology of culture past and present* (Vol. 6, pp. 1–40). Jai Press.

Le Grange, L. (2019). Ubuntu. In A. Kothari, A. Salleh, A. Escobar, F. Demaria, & A. Acosta (Eds.), *Pluriverse. A post-development dictionary* (pp. 323–326). Tulika Books.

Lien, M. (2015). *Becoming salmon: Aquaculture and the domestication of a fish*. University of California Press.

Lien, M. (2024). Fluid scalability: Frontiers and commons in salmon waterworlds. *Ethnos*, *89*(3), 401–417. https://doi.org/10.1080/00141844.2023.2213851

Lien, M., & Law, J. (2011). "Emergent Aliens": On salmon, nature and their enactment. *Ethnos*, *76*(1), 65–87.

Linder, E. (2024). *Caring for olive oil: Cultivating flows, crafts & traditions*. Stockholms universitet.

Liu, G., Sun, J., & Liu, S. (2015). Chapter 2. From fisheries toward aquaculture. In H. Yang, J. F. Hamel, & A. Mercier (Eds.), *The sea cucumber Apostichopus japonicus: History, biology and aquaculture* (pp. 25–36). Academic Press. http://dx.doi.org/10.1016/B978-0-12-799953-1.00002-7

MacDougall, D. (2006). *The corporeal image: Film, ethnography and the senses*. Princeton University Press.

References

Mark, C. K. (2017). To 'educate' Deng Xiaoping in capitalism: Thatcher's visit to China and the future of Hong Kong in 1982. *Cold War History*, *17*(2), 161–180. https://doi.org/10.1080/14682745.2015.1094058

Maulu, S., Hasimuna, O. J., Haambiya, L. H., Monde, C., Musuka, C. G., Makorwa, T. H., Munganga, B. P., Phiri, K. J., & Nsekanabo, J. D. (2021). Climate change effects on aquaculture production: Sustainability implications, mitigation and adaptations. *Front. Sustain. Food Syst.* *5*, 609097. https://doi.org/10.3389/fsufs.2021.609097

Mbembe, A. (2021). *Out of the dark night: Essays on decolonization*. Columbia University Press.

Mbembe, A., & Sarr, F. (Eds.) (2023). *To write the Africa world*. Polity Press.

Melillo, E. D. (2015). Making sea cucumbers out of Whales' teeth: Nantucket castaways and encounters of value in nineteenth-century Fiji. *Environmental History*, *20*(3), 449–474.

Menezes, A., Gueye, N., & Jolly, C. (2024). *Policies and strategic plans in support of aquaculture development in sub-Saharan Africa – A tool for planning and resource mobilization*. FAO.

Metz, T. (2023). Understanding a thing's nature. Comparing Afro-relational and Western-individualist ontologies. In P. A. Ikhane, & I. E. Ukpokolo (Eds.), *African Epistemology. Essays on being and knowledge* (pp. 63–78). Routledge.

Mmbaga, T. K., & Mgaya, Y. D. (2004). Sea cucumber fishery in Tanzania: Identifying the gaps in resource inventory and management. In A. Lovatelli, C. Conand, S. Purcell, S. Uthicke, J. F. Hamel, & A. Mercier (Eds.), *Advances in sea cucumber aquaculture and management* (pp. 193–203). FAO. Fisheries Tech Paper 463.

Mmbaga, T. K., & Mgaya, Y. D. (2007). Sea cucumbers in Tanzania. In C. Conand & N. Muthiga (Eds.), *Commercial sea cucumbers in the Western Indian Ocean* (WIOMSA Book Series No 5, pp. 51–56)..

Mohsen, M., Ismail, S., Yuan, X., Yu, Z., Lin, C., & Yang, H.. (2024). Sea cucumber physiological response to abiotic stress: Emergent contaminants and climate change. *Science of The Total Environment*, *928*, 172208, ISSN 0048-9697. https://doi.org/10.1016/j.scitotenv.2024.172208

Mol, A., Moser, I., & Pols, J. 2010. Care: Putting practice into theory. In A. Mol, I. Moser, & J. Pols (Eds.), *Care in practice: On tinkering in clinics, homes and farms* (pp. 7–26). Transcript.

Moore, J. (Ed.) (2016a). *Anthropocene or Capitalocene? Nature, history and the crisis of capitalism*. Kairos Books.

Moore, J. (2016b). The rise of cheap nature. In J. Moore (Ed.), *Anthropocene or Capitalocene? Nature, history and the crisis of capitalism* (pp. 78–115). Kairos Books.

Morningstar, N. (2023). Neoliberalism. In F. Stein (Ed.), *The open encyclopedia of anthropology*. https://doi.org/10.29164/20neolib Original work published 2003

Murphy, F., & von Roekel, E. (Eds.) (2024). *A collection of creative anthropologies. Drowning in blue light and other stories*. Palgrave Macmillan.

Mwaipopo, R., & Ndaluka, T. (2023). Local narratives on the blue economy: An analysis of livelihood mobility in coastal communities in Bagamoyo, Tanzania. *Tanzania Journal of Development Studies*, *21*(2), 1–20.

Nkrumah, K. (1963). *Africa must unite*. Heinemann.

Nonini, D. M. (2008). Is China becoming neoliberal? *Critique of Anthropology*, *28*(2), 145–176. https://doi.org/10.1177/0308275X08091364

Nyerere, J. (2011). *Freedom, non alignment and South South cooperation. A selection from speeches 1974-1999*. Oxford University Press.

References

Osseweijer, M. (2005). "We wander in our ancestors' yard": Sea cucumber gathering in Aru, Eastern Indonesia. In R. Ellen, P. Parkes, & A. Bicker (Eds.), *Indigenous environmental knowledge and its transformations: Critical anthropological perspectives* (pp. 56–78). Harwood Academic Publishers.

Palsson, G. (2013). Ensembles of biosocial relations. In T. Ingold & G. Palsson (Eds.), *Biosocial becomings: Integrating social and biological anthropology* (pp. 22–41). Cambridge University Press.

Pauwelussen, A. P. (2017). *Amphibious anthropology: Engaging with maritime worlds in Indonesia* [Doctoral dissertation]. Wageningen University and Research.

Peters, K., & Steinberg, P. (2019). The ocean in excess: Towards a more-than-wet ontology. *Dialogues in Human Geography*, *9*(3), 293–307.

Pierre, J. (2020). The racial vernaculars of development: A view from West Africa. *American Anthropologist*, *122*, 86–98. https://doi.org/10.1111/aman.13352

Puig de la Bellacasa, M. (2017). *Matters of care: Speculative ethics in more than human worlds*. University of Minnesota Press.

Purcell, S. W., Lovatelli, A., Gonzalez Wanguemert, M., Solis Marin, F. A., Samyn, Y., & Conand, C. (2023). *Commercially important sea cucumbers of the world*. FAO.

Purcell, S. W., Shea, S. K., & Gray, B. C. (2025). Decadal changes in value of dried sea cucumbers (bêche-de-mer) in Hong Kong markets. *Marine Policy*, *171*, 106450.

Qiang, Z. (2010). China's strategic relations with Africa. In A. Harneit-Sievers, S. Marks, & S. Naisu (Eds.), *Chinese and African perspectives on China in Africa* (pp. 56–69). Pambazuka Press.

Reichman, D. (2013). Entrepreneurship in a pickle: Innovation and arbitrage in the sea cucumber trade. *Anthropological Quarterly*, *86*(2), 559–588.

Rethmann, P., & Wulff, H. (Eds.). (2023). *Exceptional experiences: Engaging with jolting events in art and fieldwork*. Berghahn Books.

Richmond, M. D. (2002). *A Field Guide to the sea shores of Eastern Africa and the Western Indian Ocean Islands*. SAREC-UDSM.

Rodineliussen, R. (2024). *Underwater worlds: An ethnography of waste, pollution and marine life*. Palgrave Macmillan.

Rodney, W. (2012). *How Europe underdeveloped Africa*. Pambazuka Press. Original work published 1972

Sachithananthan, K. (1994). Beche-de-mer trade: Global perspectives. *Bull. Cent. Mar. Fish. Res. Inst*, *46*, 106–109.

Sahlins, M. (1993). Goodbye to *Triste Tropes*: Ethnography in the context of modern world history. *Journal of Modern History*, *65*(1), 1–25.

Saniotis, A. (2012). Muslims and ecology: Fostering Islamic environmental ethics. *Contemporary Islam*, *6*, 155–171.

Semesi, A. K., Mgaya, Y. D., Muruke, M. H. S., Francis, J., Mtolera, M., & Msumi, G. (1998). Coastal resources utilization and conservation issues in Bagamoyo, Tanzania. *Ambio, Building Capacity for Coastal Management*, *27*(8), 635–644.

Sheriff, A. (2010). *Dhow cultures of the Indian Ocean. Cosmopolitanism, commerce, Islam*. Hurst & Company.

Shivji, I. (2006). *Let the people speak. Tanzania down the road to neo-liberalism*. CODESRIA.

Singleton, V. (2010). Good farming control or care? In A. Mol, I. Moser, & J. Pols (Eds.), *Care in practice: On tinkering in clinics, homes and farms* (pp. 235–256). Transcript.

Smethurst, P. (2012). *Travel writing and the natural world, 1768-1840*. Palgrave Macmillan.
Standing, G. (2023). *The blue commons. Rescuing the economy of the sea*. Penguin Random House UK.
Steinberg, P., & Peters, K. (2015). Wet ontologies, fluid spaces: Giving depth to volume through oceanic thinking. *Environment and Planning D: Society and Space, 33*, 247–264.
Strang, V. (2021). Elemental powers: Water beings, nature worship and long-term trajectories in human-environmental relations. *kritisk etnografi: Swedish Journal of Anthropology, 4*(2), 15–34.
Swanson, H. A., Lien, M. E., & Ween, G. B. (Eds.) (2018). *Domestication gone wild: Politics and practices of multispecies relations*. Duke University Press.
Táíwò, O. (2022). *Elite capture: How the powerful took over identity politics (and everything else)*. Pluto Press.
Teletchea, F. (2021). Fish domestication in aquaculture: 10 unanswered questions. *Animal Frontiers, 11*(3), 87–91.
Torell, E., McNally, C., Crawford, B., & Majubwa, G. (2017). Coastal livelihood diversification as a pathway out of poverty and vulnerability: Experiences from Tanzania. *Coastal Management, 45*(3), 199–218. https://doi.org/10.1080/08920753.2017.1303718
Tsing, A. (2024). Others without history. In A. L. Tsing, J. Deger, A. Keleman Saxena, & F. Zhou (Eds.), *Field guide to the patchy Anthropocene: The new nature* (pp. 129–144). Stanford University Press.
Tsing, A. L. (2015). *Mushroom at the end of the world: On the possibility of life in capitalist ruins*. Princeton University Press.
Tsing, A. L., Deger, J., Keleman Saxena, A., & Zhou, F. (2024). *Field guide to the patchy Anthropocene: The new nature*. Stanford University Press.
Tsing, A. L., Gan, E., Bubandt, N., & Swanson, H. (Eds.) (2017). *Arts of living on a damaged planet*. University of Minnesota Press.
Uimonen, P. (2001). *Transnational.dynamics @ development.net: Internet, modernization and globalization* [Dissertation]. Stockholm University.
Uimonen, P. (2012). *Digital drama. Teaching and learning art and media in Tanzania*. Routledge. https://innovativeethnographies.net/digital-drama-teaching-and-learning-art-and-media-in-tanzania/
Uimonen, P. (2016). Digital narratives in anthropology. In H. Wulff (Ed.), *The anthropologist as writer: Genres and contexts in the 21st century* (pp. 243–253). Berghahn.
Uimonen, P. (2019). Decolonizing cosmopolitanism: An anthropological reading of Immanuel Kant and Kwame Nkrumah on the world as one. *Critique of Anthropology, 40*(1), 81–101. https://doi.org/10.1177/0308275X19840412
Uimonen, P. (2020). *Invoking Flora Nwapa. Nigerian women writers, femininity and spirituality in world literature*. Stockholm University Press. Open access at https://www.stockholmuniversitypress.se/site/books/m/10.16993/bbe/
Uimonen, P. (2023). Sacred muses: The lake goddess in Flora Nwapa's literary worldmaking. In P. Rethmann & H. Wulff (Eds.), *Exceptional experiences: Engaging with jolting events in art and fieldwork* (pp. 123–137). Berghahn Books.
Uimonen, P. (2025). Taking care of sea cucumbers: Artisanal aquaculture in the Blue Economy. *Anthropology Today, 41*(3), 7–10.
Uimonen, P., & Masimbi, H. (2021). Spiritual relationality in Swahili ocean worlds. *kritisk etnografi: Swedish Journal of Anthropology, 4*(2), 35–50.

Uimonen, P., & Rodineliussen, R. (Eds.) (2025). Caring for ocean creatures. Special issue in *Anthropology Today*, *41*(3), 3.

Weber, I. (2020). Origins of China's contested relation with neoliberalism: Economics, the World Bank and Milton Friedman at the dawn of reform. *Global Perspectives*, *1*(1), 1–14. https://doi.org/10.1525/gp.2020.12271

Westmoreland, M. (2022). Multimodality: Reshaping anthropology. *Annual Review of Anthropology*, *51*, 173–94.

Wolf, E. (1982). *Europe and the people without history*. University of California Press.

Yang, H., & Bai, Y. (2015). *Apostichopus japonicus* in the life of Chinese people. In H. Yang, J. F. Hamel, A. Mericer (Eds.), *The sea cucumber Apostichopus japonicus: History, biology and aquaculture* (pp. 1–35). Elsevier Science & Technology.

Yang, H., Hamel, J.-F., & Mericer, A. (Eds.) (2015). *The sea cucumber Apostichopus japonicus: His, biology and aquaculture*. Elsevier Science & Technology.

Index

Note:- Page references in *italics* denote figures and with "n" endnotes.

active buoyancy adjustment (ABA) 54
adaptation and growth 65
Advances in Marine Biology 29
Akolor, Kodjo 39
algae 13, 71, 74, 80
Ammar, N. 35
animals 8, 38, 61, 89; classification of 60; cognition 56–59; domestic 67; exotic 89, 91; farming 70; human interactions with 26; husbandry practices 66; wild 70
Animal Welfare Bill 60
Anthropocene 8, 9, 27, 105
anthropology 4–8, 16, 19, 20, 24–27, 33, 34
Apostichopus japonicus 100
aquabiopolitics 83–84
aquaculture 97–101; artisanal 2, 67–70, 81; climate change impacts on 85; industrial 71, 110; known risks in 83, 85; salmon 8, 67, 69, 70, 100, 103; *see also* fish/fishing
aquatic food 110
articulation 26, 30, 33
artificial intelligence (AI) 4–5, 18, 39, 65, 67; challenges in Capitalocene 89; chatbot 4, 5, 19; generated story 19; imagination 4–5, 44; *see also* chatGPT
artisanal aquaculture 2, 67–70, 81; *see also* aquaculture
Arusha Declaration in 1967 98
avoiding strong currents 45
Azania 90

Bandung Conference in 1955 98
Baraka 51–52, 83
barefoot methods 10–15

Battuta, Ibn 90
Beach Management Unit (BMU) 13, 77, 102
beche-de-mer (trepang) 9, 24, 34, 89, 91; in anthropology 24–27; import of 92; postcolonial exploits of 92–93; selling 25; trade 20, 30, 89
Becoming Salmon (Lien) 59
behavior and habitat suitability 64
biological rhythms 45
Bioluminescent plankton 6
biosecurity 83–84
blue economy 4, 9, 20, 69, 70, 77, 89, 101, 102, 110
blue expansion 87–107
Bohadschia 32
Book of Waves (Helmreich) 7
Brummett, Randall E. 110
Bubandt, N. 15, 26

capitalism: global 9, 62, 97; history of 26, 30; racial 27, 33, 101; scientific 100–102
Capitalocene 9, 20, 27, 91, 111; artificial intelligence (AI) challenges in 89; early history of 30–34; survival in 105–106
Caribbean island 15
CCTV cameras 72, *73*
Challenger Expedition 28
ChatGPT 5, 89; *see also* artificial intelligence (AI); OpenAI
Chinese cuisine 108
Chinese culture 38, 41–42, 89, 108
Chinese traders 91, 95, 96
Chou Enlai 98
clandestine fishing 97

climate change 20, 67, 82–86, 105, 106, 110
Commercially Important Sea Cucumbers in the World 108
coral reef 15, 29, 106
crabs 13, 15, 21, 45, 50, 58, 60, 64, 74, 83

da Gama, Vasco 91
Darwin, Charles 28
De Holothuriis (Jaeger) 31
Deng Xiaoping 97–99
disease risks 65
dispossession 101–105
disruptions to ecological roles 106
domestication 64–86; characteristic of 76; culmination of 80; partial 20, 70, 76; of salmon 83

Echinoderms 27–30
ecology of visual inscriptions (Grasseni) 31
elite capture 101–105
environmental stressors 64, 85
environment-making 27, 34
Escobar, A. 61
ethical considerations 65
Europe and the People without History 24

Fanon, F. 103
farm/farming 3, 47, 67, 95 *66*; divers filming sea cucumbers *17*; group interview 10, *10*; Kaole 16, *22*, 49, 76, 77; multispecies and multimaterial *14*, 70–75; in ocean 64–67
feeding practices 42, 65
Ferguson, J. 100
A Field Guide to the Seashores of Eastern Africa and the Western Indian Ocean Islands (Richmond) 28, 31, 37, 56
fisheries 77, 85, 86, 105, 106, 109–110
The Fisheries Education and Training Agency (FETA) 51–53, 69, 77, 82, 83, 95
fish/fishing 11, 46; clandestine 97; community 13, 49; domestication 70; illegal, unreported and unregulated (IUU) 105; jellyfish 74; net 46; pearlfish 58, 63n2; salmon (*see* salmon fish); tilapia fish 78; and traders 93–97; wooden 90; *see also* aquaculture
Food and Agriculture Organization (FAO) 78, 99–100, 109, 110

Gaga, Lady 39
global capitalism 9, 62, 97
Global Knowledge on the Commercial Sea Cucumber Holothuria Scabra (Hamel) 29, 37
God's creation 34–36
Gogo Hotel beach camp 94
GoPro camera 16, *17*, 18, 19, 82
Gruber, M. 18

habitat destruction 106
handling stress 65
Haraway, D. 4, 7–9, 23, 36, 61
Hayles, Kathryn 57
health and disease 65
heaven and earth 40–41
Helmreich, S. 7
Hidaya cyclone 84
Hofmeyr, I. 7
Holothurians 27–30
Holothuria scabra (sandfish) 29, 42–43, 72, 93; adaptation and growth 65; behavior of 53–56, 64; collection 31; danger lurks 42; drifting 63; environmental stressors 64; ethical considerations 65; farming in Ocean Pens 64–67; feeding 42; Great Ocean Cycle 43; health and disease 65; human interaction 65; IUCN listed as endangered 93; morning on the seabed 42; night on the seabed 43; reproduction of 43, 55, 62–63; scientific portrayal of 109; social life 43; whispers of 86
human interaction 65
hybrid communities 58

illegal, unreported and unregulated (IUU) fishing 105
Indian Ocean 6–9, 25, 85, 91, 100, 109
Indo-Pacific region 9, 28, 30, 53
industrial aquaculture 71, 110; *see also* aquaculture
industrial production 78–82
industrial revolution 92
Ingold, T. 8, 38, 57–59, 61
inscription process 30, 33
Institute of Marine Science (IMS) 83
International Monetary Fund (IMF) 98, 101
IUCN Red List 93, 109

Jaeger, Wilhelm Friedrich 31
jellyfish 74

Index 123

Jongoo Bahari (Sea Cucumber) 1, 18, 34–36
Jue, M. 7

Kaiza, Victor 95
Kant, Immanuel 34
Kaole farm 16, 22, 49, 76, 77
Khaldûn, I. 57
Khatib, Mary 16
Kirksey, E. 7

Latour, B. 9, 30–31, 33
Lazard, Jérôme 110
Lestel, Dominique 58
Lien, M. 8, 59, 61, 66–67, 71, 83, 103
Linder, Elin 23
Linneaus, Carl 28, 30, 34
Living with the Ocean 18
lobsters 11, *12*, 36, 60
London School of Economics and Political Science (LSE) 60

MacDougall, D. 16
Madagascar 68, 80, 83, 93, 101
Mafia Island 16, 90, 92
manual maintenance work *73*
marine ecosystem 27, 33, 82, 89, 106, 112
marine hatchery 78–82
marine science 19, 27–30, 53–56, 58
Masimbi, Hussein 11, 16, 79
Mbembe, A. 57
Melillo, E. D. 25–27, 34
Mercedes-Benz cars 107n3
Metz, T. 60
microplastic and heavy metal pollution 106
Moehl, John 110
moisture retention 45
The Monkeys and the Sea Cucumbers 40
Moore, J. 27, 92
Mtwara 16, 51–52, 69, 76–77, 82, 83, 94, 95, 111
multimodal/multisensory methods 16–18
multispecies: correspondence 56–59; ethnography 4, 7, 8, 16, 22; and multimaterial ocean farms *14*, 70–75; ocean worlds 10–15; storytelling 4–6; worldings 8, 24
The Muqaddimah (Khaldûn) 57

natural habitat mimicry 64
natural science 33
neocolonialism 103

neoliberal expansion 97–101
net fishing 46
Nkrumah, K. 103
non-governmental organization (NGOs) 103
Norwegian Agency for Development Cooperation (Norad) 109
Nyerere, Julius 98

ocean: acidification 106; cleaners 1, 51; creatures 108–112; farmers 67, 69; grabbing 9, 89, 102–104; *Holothuria Scabra* in 64–67; sea cucumbers protecting 112; space 74, 101, 103–105; as theory machine 7; warming 106
octopus 6, 60, 94
Omani rule 91
ontological occupation 61
OpenAI 5; *see also* ChatGPT
Order Holothuriida 37
Osseweijer, M. 15
overexploitation and illegal harvesting 105

partial domestication 20, 70, 76
pearlfish 58, 63n2
People's Republic of China 92
Peters, K. 7
Pinochet, Augusto 97
plantation economies 105
plants and specimen 28
Polo, Marco 90
Portuguese Empire 91
predation 64
private-public-partnership (PPP) 81
Public Service Reform Programme 99

racial capitalism 27, 33, 35, 101
Reagan, Ronald 97
Reichman, D. 25
reproduction 20, 38, 43, 49, 50, 55, 56, 62–63, 65–67, 76–77, 80, 81
rest and recuperation 45
Rodineliussen, Rasmus 63n1
Roman Empire 90

salmon fish: aquaculture 8, 67, 69, 70, 100, 103; cultivation of 76; domestication 69, 83; farms 71, 83
saponins 58
Scientific Benefits of Sea Cucumber 50
scientific capitalism 100–102
scientific gaze 30–34

sea cucumber: aquaculture 97–101; association 72; barefoot methods for 10–15; behavior in coastal Tanzania 46–53; being dried for export 88; burrowed into sand 46; business 97; in Chinese Culture 41–42; dead 85; fishers and traders 93–97; in group interview 10, 10; and human becomings 67–70; imagination and storytelling with AI 4–5; introduction 1–20; lobsters and 12, 26; message from 112; mobility 55; monkeys and 40; narrative structure 18–20; ocean environment 111; raw and processed 48; relational worldmaking 2; reproduction of 81; scientific classification of 28; seafloor 2; species kinds and relational becomings 36–38; spitting water 23; trade 9, 24, 25, 89, 91, 92, 99; see also farm/farming; stories "Sea cucumbers: Southern Tanzania's marine gold" (Kaiza) 95
sea products 96
sea snail 50
seawater environment 74
Second World War 26, 91
sediment degradation 65
sediment stability and feeding 45
sensory worldings 44–63; introduction 44–46; and sentient becomings 59–62; see also Holothuria scabra (sandfish)
sexual reproduction 55, 56; see also reproduction
shark fins 92
Sjögurka (Akolor) 39
Skin Ulceration Disease (SKUD) 83
small-scale farmers 85
snorkeling method 15
Solomon Islands 54, 56, 109
space and movement 64
species kinds and relational becomings 36–38
A Star is Born (Gaga) 39
Steinberg, P. 7

stories: with AI 4–5; multispecies 4–6; prompts for 9; reproduction 62–63; sea cucumber *(sea ginseng)* 38–42, 108–111
suffering domestication *see* domestication
Swahili coast 1, 3, 24, 34, 35, 87, 88, 90, 91
Swahili mythology 40–41
Swahili Ocean Worlds 3, 4, 11
symbiotic multispecies community 58
Systema Naturae (Linneaus) 28, 31

Tale of the Golden Sea Cucumber 87–89
Tanzania 6, 34, 55, 71, 76, 93, 98, 101
TaSUBa arts college 18
temperature and salinity 64
Thatcher, Margaret 97, 99
Thelenota Ananas 31
tilapia fish 78
transoceanic trade, history of 89–92
Trepang Ananas 31, 32
Tsing, A. L. 26, 33, 44

underwater creatures 21–43; beche-demer 24–27; in Capitalocene 30–34; Echinoderms and Holothurians 27–30; as God's creation 34–36; overview 21–24

Wabenzi 107n3
walking on seafloor 11, 14
Wallace, A. R. 33
water quality 64
Western Indian Ocean (WIO) 28
wet ontologies 7
WhatsApp 82
When Species Meet (Haraway) 7, 36
wild and farmed fingerlings 76–77
Wolf, E. 24–27, 91
wooden fishing 90
World Bank 98, 101

Yusuf, Abdillahi 95–96

Zanzibar 50, 52, 67, 91, 103, 111
Zanzibar Marine Hatchery 78, 79